STAR GAZERS

STAR GAZERS

Finding Joy in the Night Sky

DAVID H. LEVY

THE UNIVERSITY OF
ARIZONA PRESS
TUCSON

The University of Arizona Press
www.uapress.arizona.edu

We respectfully acknowledge the University of Arizona is on the land and territories of Indigenous peoples. Today, Arizona is home to twenty-two federally recognized tribes, with Tucson being home to the O'odham and the Yaqui. Committed to diversity and inclusion, the University strives to build sustainable relationships with sovereign Native Nations and Indigenous communities through education offerings, partnerships, and community service.

ISBN-13: 978-0-8165-5464-5 (paperback)
ISBN-13: 978-0-8165-5465-2 (ebook)

Cover design by Leigh McDonald
Designed and typeset by Sara Thaxton in 10.5/14 Warnock Pro with Atrament

Unless otherwise noted, all photos are courtesy of the author.

Library of Congress Cataloging-in-Publication Data
Names: Levy, David H., 1948– author.
Title: Star gazers : finding joy in the night sky / David H. Levy.
Description: Tucson : University of Arizona Press, 2025. | Includes bibliographical references and index.
Identifiers: LCCN 2024032682 (print) | LCCN 2024032683 (ebook) | ISBN 9780816554645 (paperback) | ISBN 9780816554652 (ebook)
Subjects: LCSH: Astronomy—Popular works. | LCGFT: Essays.
Classification: LCC QB44.3 .L485 2025 (print) | LCC QB44.3 (ebook) | DDC 520—dc23/eng/20241119
LC record available at https://lccn.loc.gov/2024032682
LC ebook record available at https://lccn.loc.gov/2024032683

Printed in the United States of America
♾ This paper meets the requirements of ANSI/NISO Z39.48-1992 (Permanence of Paper).

To my family and closest friends,

particularly for David and Pam Rossetter,

who have devoted so much time to making certain I was all right during these difficult years after I lost sweet Wendee,

I dedicate this book.

This picture was taken by Joyce Stein, from her backyard, on April 8, 2024. Assistance was provided by Dr. Lawrence Stein. Venus can be spotted between the telephone wires.

CONTENTS

STAR GAZERS

PART ONE

NOT HOW TO WATCH THE SKY, BUT WHY

I shall never forget an incident in my life that occurred sometime during the 1970s. I was enjoying an evening chat with my dad about reading. Dad loved to read. Furthermore, like President Kennedy, he devoured books with his ability to read rapidly. On that particular evening, I brought him a copy of Leslie Peltier's autobiography *Starlight Nights: The Adventures of a Star-gazer* (Harper and Row, 1965). As he held the book carefully, he inquired about why I had read and reread it, and I told him that there are many books that tell the reader how to observe the night sky, but this one explains why one should embrace the night, and the stars.

About two days later, Dad returned the book. "David," he intoned, "you probably remember how, when you were much younger, I thought you were spending too much of your time cogitating about the stars. I was wrong. Especially since, at that time, I had no notion of what you would accomplish with this passion. Eventually I came to understand and appreciate your passion. Since then, I have followed your stargazing with interest. Now, after reading this wonderful book, I have a better idea of why you, my son, are so entranced with stargazing."

This book strives to achieve that same goal. I think that the best way to accomplish that is to explore the stories, failures, goings-on, activities, events, and thinking of a selection of people whose lives have been ingrained upon my own. Not all these people are astronomers. What unites them is not the sky itself, but their ways and feelings about the sky. Moreover, not all these people are people. Some are telescopes. Some are constellations. Some are comets. Some are celestial events like eclipses. And some are me. Separately, each has something to say. Together, they paint a picture of what it is like, with some thought and even a little naïveté, to launch oneself from the trivia of the daily broadcasts and examine the big picture, the cosmic picture, found by simply looking up.

CHAPTER 1

STAR GAZERS

I open this book with Wordsworth's poem "Star-Gazers." It succinctly makes the point that many people can be disappointed with what they see through a telescope unless what they see is as spectacular as what the James Webb Space Telescope offers. Not I. My best views come through telescopes I actually get to look through.

The title of this piece, "Star Gazers," comes from the 1806 William Wordsworth poem. That poem is about a telescope "upon its frame, and pointed to the sky" and whose observers, after paying a fee, get a look at the cosmos. But the heart of this poem is that they are dissatisfied with their view. They refuse to believe what they see, or they expect a view similar to those observed through the largest telescopes available, or they simply do not understand. That feeling persists to this day. About thirty years ago, I encountered a family that had just looked through the 16-inch diameter telescope at our local Flandrau Planetarium at the University of Arizona in Tucson. The father was complaining about the telescope's poor optics. Having looked through that telescope myself, I knew that the optics were fine. It seemed he was taking a line right out of the poem.

Wendee, in 1996, with Echo, my first telescope.

A few years after that, I lent a small apochromatic refractor, named Surveyor after the unmanned lunar program active during the mid 1960s, to be mounted atop the 16-inch reflector at the Flandrau Science Center. It was successful for many years, but it now exists, still providing exquisite views of the sky, at the Jarnac Observatory.

As a poem, "Star-Gazers" can be interpreted as satire. Like the book which shares its title, its basic idea is not really how we should use a telescope to watch the stars, but why such an instrument can augment our sense of wonder. As we look across space and back through time upon the night sky, we are whisked away from the specifics of our lives and ushered instead into the big picture of the cosmos.

* * *

> What crowd is this? What have we here? We must not pass it by;
> A telescope upon its frame, and pointed to the sky . . .
>   —WILLIAM WORDSWORTH, "STAR-GAZERS"

While I was working on my master's degree at Queen's University in Canada in the mid 1970s, I came across this poem, loved it, and decided to include it in my thesis. Norman MacKenzie, my thesis advisor, a scholar and a genius, penciled one comment at the bottom of this poem: "Wordsworth wrote some wretched verse." Norman did not have much of a sense of humor, but this was brilliantly funny.

In his poem, Wordsworth complains about how many people who look through a telescope are disappointed in what they see. At no point in time is that idea more cogent than now. If a telescope we look through cannot offer us a view as good as a space telescope's, then that telescope is a failure.

By the end of the poem, the crowd abandons the telescope:

One after one they take their turns, nor have I one espied
That doth not slackly go away, as if dissatisfied.

For me, the night sky is far more than our imagined perceptions of what we can see through a telescope. Some of us can look at an internet photograph all day long, but not even try to look at the same object through a telescope. The beauty of the sky lies in its reality. The planets I see are real worlds. The constellations I point out to young observers contain real stars. One evening I asked a group if they had seen the recent eclipse of the Moon. "Yes," answered one, "I saw it online." No, he didn't. Eclipses are real only if you actually see them in the sky, while they are happening.

It is a given that a backyard telescope will never show us Jupiter as detailed or as colorful as a telescope out in space will. What that telescope does show us is the genuine sky, a sky without artificial color enhancement, a sky as it really exists on top of our heads on every clear night. It shows us a sky untarnished by the trivial events of the day, and unspoiled by petty concerns that are bothering us. Our own telescope truly shows us the Moon as it was a third of a second ago, a star as it appeared thirty-four years ago, or a galaxy as it appeared twelve million years in the past. Our backyard telescope shows us what is there, and, unlike the crowd from 1806 that left dissatisfied, the people of today can understand that the sky they see is real.

—March 2022

CHAPTER 2

OMICRON!

If the previous chapter explores spacetime, the next one focuses on the meaning of those who have shared the cosmos with me over the decades. We have room for personal loss here. Losing my friend Don Machholz, the world's foremost visual discoverer of comets, was particularly sad. But the more recent forfeiture of my closest childhood friend, Carl Jorgensen, from COVID-19 is particularly hard to accept. We became friends at the observatory of the Royal Astronomical Society of Canada's Montreal Centre, in the fall of 1963. Isabel Williamson, the director of the group's observing program, said, "Young Carl Jorgensen, meet young David Levy." Despite these sad moments, this segment celebrates their memory not through them but through the sky they loved. As I particularly remember Carl, and the double stars he loved and encouraged me to observe, I focus now on one of my very favorite multiple stars, a nearby triple sun called Omicron² Eridani. The star is fabulous to look at, but as we shall see its importance transcends the real night sky into the wondrous, though fictitious, world of Star Trek.

* * *

You must have read dozens of articles, online or in print, about the Omicron variant of COVID-19 which raged across the year 2022. Fortunately, this is not one of them. This article is about Omicron^2 Eridani. It is a faint star in the constellation of Eridanus, the River.

Actually, there are two Omicron stars in that constellation. The first is brighter and is a variable star. The second one is one of the closest stars to the Sun. Omicron^2, also known as 40 Eridani, happens to be not a disease but one of the most interesting star systems in the entire sky.

Omicron^2 is a triple star system that is only about sixteen light years away. Its brightest component is a Sun-like star faintly visible to the unaided eye on a good night. It lies in northern Eridanus, the River, just a few degrees west of Rigel at Orion's right foot. The secondary star is a white dwarf star. Unlike the companion of Sirius, this star is ninth magnitude and not near the brighter star, so it is easy to see in a small telescope. The tertiary, or third star, is not far from the secondary, but at eleventh magnitude it is also not difficult to spot. This third star is a red dwarf.

Although red dwarf stars are the most plentiful, by far, in our region of the Milky Way galaxy, they are almost impossible to see because they are so small. The closest one to us is Proxima Centauri, or Alpha Centauri C, which at 4.24 light years distant is the closest star to the Sun. Also, because they are so small and intrinsically faint, only a few of them are easy to find. The third star of this lovely triple, 40 Eridani C, is one of the easiest to find.

This interesting star has something else going for it. In 2018, astronomers discovered a planet orbiting the primary star. With a rapid and close orbit around Omicron^2, such a planet would receive much more radiation from the primary star than Earth gets from the Sun. But in 2021 new observations cast doubt on whether this planet exists at all.

Whether Omicron^2 Eridani really hosts a planet is subject to debate. But in the universe of *Star Trek*, it surely does. It is the star of Vulcan, Mr. Spock's home world. In the episode "Operation—Annihilate!," which appears near the end of the first season, Spock is blinded by the intense light used to immobilize the invading parasites on the planet Deneva. However, his blindness is temporary because of the existence of an inner eyelid. Vulcan is said to orbit Omicron^2 Eridani's primary star, and, since it is so much brighter than our Sun, even though Vulcan is at the same distance that Earth is from our Sun, Vulcans need the inner eyelid to protect their eyes.

I rather enjoy the idea that the fictitious Vulcan happens to orbit one of my favorite real stars. We admire Omicron^2 Eridani, the real star, and wish it will "Live long and prosper."

—April 2022

CHAPTER 3

A STREAK OF LIGHT

I have several stories to tell of how I got interested in the night sky, but the shortest and most direct one is of a shooting star I saw on the evening of July 4, 1956. This goes back a very long way. From the date of this shooting star's flash across a corner of the sky to the time I hope that this book appears, almost seventy years have passed. Locked in my memory, this event proves that the hobby of stargazing can either be very expensive, or it does not have to cost a dime. And even after all these years, watching shooting stars remains one of my happiest pastimes. This particular aspect of astronomy is also the cheapest. Using just your eyes and relaxing on a comfortable lawn chair, just gaze at the sky, learn it on your own terms, and watch for shooting stars as they fall.

★ ★ ★

It was a flash, a single streak of light, that got me started in astronomy more than sixty years ago. The streak could not have lasted more than a second that clear evening of July 4, 1956. I was terribly homesick. At age eight, just four days into my first summer away from home, I had already written to beg Mom and Dad to rescue me from that lonely place. I did not understand at the

An Orionid meteor streaks in front of the Andromeda galaxy.

time that they needed a break from me and that, no matter what happened, I wasn't going home until the end of the summer. The sky was clear that warm summer evening as children and staff gathered around the softball field to enjoy a fireworks display in celebration of the Fourth of July. As a young Canadian, I didn't know anything about what the United States Independence Day stood for.

As the fireworks wound down, the youngest groups, including mine, were dismissed for the night. We began walking up the hill toward Bunk B. As we strolled up the hill, my glance accidentally turned toward the darkening sky above me. Stars were coming out. I saw one bright star high in the east, and several fainter stars around it. It was beautiful, though I had no idea yet what this beauty would eventually mean to me. I just gazed upward.

Then it happened. A streak of light scratched the sky, flying toward that brightest star. Startled, I asked the others if they had seen it, too. Since none of them had been looking upward, they all said no. Interestingly, none of the other children teased or made fun of my observation or me. Far ahead of its time, this particular camp had no place for bullying, and all the children were treated with respect. I looked again at the sky. Is it possible, I thought, that this shooting star was meant just for me? I simply placed that little memory in my eight-year-old brain where it rested for about a year until October 4, 1957. I recalled it when I was told that the Russians had launched a rocket into orbit around Earth. To me, that dawn of the Space Age was intensely private because I could relate it to something I had personally seen.

The image of the meteor rested until June 1960, when a bicycle accident and a get-well present of a book about astronomy brought the memory to the forefront of my mind. This time it stayed there. This time I was hooked for life. I know now that my first meteor was from the Omicron Draconid meteor shower, an annual event confirmed at about that same time by a young astronomer named Brian Marsden. It is likely that my shooting star was the first confirmed visual sighting of an Omicron Draconid meteor. I've seen more since, and on July 4, 2005, I photographed one that happened to be passing in front of comet Tempel just minutes after the Deep Impact spacecraft crashed into the comet. Over the next several decades I saw thousands more meteors, particularly on a late November night in 2001 when I counted no fewer than 2,406 meteors during the Leonid meteor storm that year.

I'll never forget that distant night, at the dawn of my life, when I saw my first shooting star that ignited a lifetime passion for the night sky.

—November 2016

CHAPTER 4

THE METEOR SHOWER THAT WASN'T, BUT NOT SO MUCH

Over the course of my lifetime, I have seen more than three hundred meteor showers, from almost unidentifiable single meteors that appear for a fraction of a second and then vanish, to the unforgettable night of November 18, 2001, during which Wendee and I counted the falls of 2,406 meteors. Except for the cost of a voyage to Australia, watching this meteor shower cost us nothing. No telescope was needed, nor was any other equipment necessary. Meteors are best seen with the unaided eye.

I have already explained how my lifelong romance with the night sky began with the sighting of a single meteor. I recall that moment each time I see a fresh shooting star appear in the night sky. And, thus, naturally I was most intrigued when I read about a possible eruption of meteors in 2022.

★ ★ ★

On May 30, 2022, observers all across the Western Hemisphere were outside, hoping to see a wonderful "new" meteor shower. The shower is actually not new. It is called the Tau Herculids, and it sends us dust particles spread around the orbit of comet Schwassmann–Wachman 3. In 1995, this normally faint comet

Tau Herculid meteor, bright, May 30, 2022.

brightened dramatically as it split into several parts, releasing huge amounts of dust into space.

At 10 p.m. mountain standard time that evening, Earth plowed through the debris released in 1995 when the parent comet broke apart. We were hoping for a possible meteor storm of hundreds of thousands of meteors. Wendee and I sat outside at Jarnac Observatory, waited, watched, and waited some more. There was one bright meteor that seemed too far from the direction my camera was pointing for its lens to detect. Ten o'clock came and went, and we counted a few shooting stars here and there. Over the course of the evening we counted eighteen meteors. But a meteor storm? To use the Yiddish word that means what you think it means, we saw bupkes. Somewhat disappointed, we went indoors and completed a quiet evening.

The next day, I examined the pictures I took. I have found that it is very difficult for a camera to record all but the brightest meteors, even from the major showers. But the second picture I saw captured the bright meteor I saw just south of Corvus in

Hydra, and the third frame recorded a fainter one. All in all, the camera counted five meteors, only the first of which I actually saw. And one frame displayed two meteors!

Even though these meteors were generally faint, they moved so slowly that they showed up nicely on the camera. So, this crazy little shower produced more meteors on camera than any other meteor shower I have witnessed. The experience proved to me that meteor showers, while poorly predictable, do offer surprises, and this one certainly did.

There was more. In Electronic Telegram 5125 of the International Astronomical Union, Daniel Green suggested that "a very faint glow from scattered sunlight may be visible in the sky centered . . . in Leo." I had no difficulty at all seeing that glow in Leo, particularly when I used averted vision, and I also noted its absence on the following night. (I saw a similar glow during the strong Perseid meteor shower in 1992.)

The best (by far) meteor shower I saw was the Leonids, from near Alice Springs, Australia, in 2001. During that night, Wendee and I counted 2,406 meteors. This year's Tau Herculids might have been less than stellar, but the sky was clear, the night was beautiful, and we enjoyed being outside as planet Earth raced through the emptiness of space, picking up cosmic dust on its windshield along the way.

—July 2022

CHAPTER 5

THE ADIRONDACK ASTRONOMY RETREAT

The Adirondack Astronomy Retreat is one of the most success-ful and most fun things that Wendee and I ever arranged. It be-gan when I started corresponding with a Mr. Bill Ehmann from The State University of New York at Plattsburgh. After the tour of the beautiful Twin Valleys campsite in 2003, he was driving me northwards back to Plattsburgh. He looked toward me and asked, "Any ideas?" I looked back at him and said, "How about an astronomy retreat here each summer?" He looked at me, smiled, and nodded. That is how the AAR was born.

★ ★ ★

The Adirondack Astronomy Retreat (AAR) ran in the Adiron-dack Mountains near Lewis, New York, from 2004 to 2019 un-der the directorship of Wendee and me. In 2019 we had a special program with lectures, a banquet featuring, among other VIPs, my brother Gerry and his partner Duane, and President John Ettling of SUNY Plattsburgh. We even presented to Dr. Ettling the first Starlight Night Prize to celebrate the university's commitment to keep this wonderful place as dark as possible. We concluded the week by burying a time capsule.

Much as we tried, the enthusiasm for the event was too strong just to end it. Now, under the direction of Patrice Scattolin from Montreal and his family, AAR is continuing. With his high intelligence and brilliant sense of humor, Patrice has run the event with an efficiency and alacrity rarely seen. Laurie Williams, with the assistance of daughters Clara and Sophie and son Marc, has kept the indoor portion running smoothly. And this year the weather helped "big-time." We had four beautiful nights, and good portions of two others. Using the camp's Meade 14-inch Schmidt–Cassegrain called Aart, a 26-inch reflector dubbed Enterprise, and Carl Jorgensen's 8-inch reflector named Pegasus, I did almost twenty-five hours of visual comet hunting. This total is possibly a record for this site. When the sky is at its best here, I can glimpse Messier 33 with the naked eye, and I did that almost every night. The International Space Station made a nice pass, and we saw several bright meteors heralding the onset of the Perseid meteor shower.

The purpose of this particular retreat was and still is to recharge our astronomical batteries, and to remind us why we became amateur astronomers in the first place. While some years let us have plenty of downtime to enjoy movies and singalongs, in other summers the night sky occupied pretty much all our time. It was truly spectacular.

While the site may be superb now, we chose it for our star party because of the memories that flood back every time I revisit it. It provided my first serious dark-sky experience decades ago, during the summers of 1964, 1965, and 1966. I loved it so much back then that I asked Dad if I could attend SUNY Plattsburgh the rest of the year. In one of the few mistakes Dad ever made, he resisted, preferring that I attend Montreal's McGill University instead. I flunked out of McGill twice. But I have never forgotten the pristine beauty of SUNY Plattsburgh's Twin Valleys campsite, with its unparalleled views of the "forever wild" Adirondack Mountains. May this priceless spot continue to remind future generations of how beautiful the mountains are, and how beautiful the night sky remains far above their lofty peaks.

—October 2019

CHAPTER 6

GETTING LOOSE CHANGE

Thus far, this book has devoted its words and pages to the night sky of our own time. What follows is a journey back through time. Instead of a comet seen last month, or last year, we can visualize a sky that would have been visible in Shakespeare's time. In Julius Caesar he not only reviews the assassination of the Roman leader but also romanticizes it with Calpurnia's prediction of a great comet.

Somewhere in the sky beyond the most distant worlds, Caesar's comet still lurks. At the time it was also known as the Julian star (Sidus Iulium), and as The Star of Caesar (Caesaris Astrum). A 1997 study by Ramsey and Licht (University of Illinois at Chicago) provides some evidence about the comet's spectacular outburst by six or seven magnitudes in July 44 BCE. Brian Marsden, who wrote the book's foreword, suggests that the very existence of this ancient comet cannot be confirmed.

★ ★ ★

More than two thousand years ago, getting loose change was about as easy as it is today. Hand a shopkeeper a silver dollar in today's world, and you can expect four quarters in change. What isn't the same as today is the design of the coin one might want to get change for. Hand the same shopkeeper a Roman coin from the first century, especially one with a bright comet engraved on its head, and one of two things might happen. Either you'd get thrown out of the store, or the shopkeeper would treat you to dinner and then bequeath his children to you. After all, if the shopkeeper read Shakespeare, he would know that the coin was celebrating Julius Caesar's Great Comet, the comet that appeared in the northern sky during the games held shortly after the assassination of Julius Caesar on the Ides of March, March 15, 44 BCE. In Shakespeare's tragedy *Julius Caesar*, Calpurnia even predicts the murder, and the comet:

> When beggars die, there are no comets seen;
> The heavens themselves blaze forth the death of princes.

In Shakespeare's play, Caesar was assassinated on the Ides of March, 44 BCE. The play mentions neither the games nor that they were played in celebration of the new emperor, Augustus Caesar. A bright comet was apparently visible in the northern sky during those games. It was widely interpreted as Julius Caesar's soul on its way to the stars. At the time, comets were omens. Calpurnia was well aware that her husband's death could be preceded or followed by a bright comet. And decades later, Seneca, in his anxiety to avoid execution by the suspicious Emperor Nero, insisted that the bright comet of AD 61 was a favorable omen to Nero. (It didn't work; Nero had Seneca put to death four years later.)

To engrave a comet on a coin may seem strange, but in fact most people never get to see a bright comet, an apparition in the night with a head and flowing tail, in their entire lives. I have. My nights under the stars have been brightened by the light of more than two hundred comets. Only a few of these comets were visible without the aid of a telescope, and most were only barely seen as specks of slowly moving haze. But even these were magical.

Comets have appeared in literature all over the world, in almost all languages, because writers since time began have seen comets and have become fascinated by them; writers like Geoffrey Chaucer, like Alfred, Lord Tennyson, like James Joyce, and like me. I caught the comet bug when I was twelve years old. Our teacher in the sixth grade, Mr. Powter, wanted us to give speeches. The topic I chose was comets. I was interested in their appearances in the sky, their appearances in history, in art, and in literature. What I knew nothing about was their role in the origin of life on Earth. I was far too young to consider the possibility that when comets collided with Earth, their debris included the CHON particles—carbon, hydrogen, oxygen, and nitrogen—the alphabet of life. Thirty-four years later, one of the comets I helped discover taught me that lesson as it careened into Jupiter in one of the major events in the history of science. This comet didn't get onto a Roman coin, or even a modern one, but it did find its way onto a German stamp. Not too bad for a tiny comet that wandered through the solar system for eons, gradually got attracted into an orbit about Jupiter, and then, in a series of explosions, reconstructed our understanding of how life could begin on a world.

—December 2018

PART TWO

TELESCOPES

Using a telescope—any telescope—will add enormously to anyone's enjoyment of the sky at night. For me, looking through my first telescope, more than sixty years ago on September 1, 1960, brought me a wealth of riches that have thrilled my heart and my mind ever since. I may in fact be somewhat naïve when I write these words, but I have never looked through a truly poor telescope.

The next few chapters will look at telescopes from different points of view, from my own first telescope, all the way forward to the new James Webb Space Telescope. Each telescope offers its own magic, its own interpretation of what the sky has to offer. Some telescopes I have used are from my own Jarnac Observatory; others are from other people's homes or star party sites. The chapter includes a look at Pegasus, the highest-quality telescope I own. The stellar views through it are simply miraculous.

CHAPTER 7

A FAVORITE TELESCOPE

My favorite telescope has got to be the one I first looked through, on September 1, 1960. I pointed it to the brightest thing in the sky. When I peered through it, the planet Jupiter waved back, along with his bright moons. I shall never forget that night, and I like to think that Mom and Dad, from wherever they are now, will not forget it either.

In 1979, when my parents sold our childhood home on Upper Belmont Avenue, I was watching television with Mom when she told me that she had found some earlier writings of mine stored in the attic. It turned out that they were some of my earliest observing records. Because I thought these records were forever lost, I was thrilled beyond expression. These records are now housed at the Linda Hall Library in Kansas City.

*   *   *

What is your favorite telescope? What is the best telescope in the world to use while viewing the night sky? For most of us, I believe the answer would be the James Webb Space Telescope, or perhaps, for older readers, the Hubble Space Telescope. Since its launch in 1990, and its successful repair in 1993, Hubble (and

Echo, my first and favorite telescope.

now Webb) has provided the majority of the beautiful images of stars, galaxies, clusters of galaxies, and almost everything we find wonderful to gaze at in the night sky. I am not an exception to this general rule: the images that Hubble returned to Earth of the impacts of the comet Shoemaker–Levy 9 during its collision with Jupiter in 1994 were absolutely breathtaking. I cannot deny that. And soon, the Vera C. Rubin Observatory, with an eight-meter diameter mirror and a field of view covering six Moon diameters, will send us pictures of a sky we know almost nothing about. So, shouldn't one of these be my favorite telescope? They surely would be, if only I could have looked through them. Neither the Rubin, nor Hubble, nor Webb is my favorite. If one of them isn't, then what is? My own favorite is a tiny 3.5-inch diameter telescope with a black tube that I've owned for more than half a century. Echo (this telescope's name) does not have the aperture or the power to spot anything other than the brightest objects in the sky. But because it gave me my very first telescopic view of Jupiter and inspired me to go into astronomy, it therefore is my favorite telescope.

I have had my eye on the sky since the warm, clear evening of July 4, 1956, when I saw a shooting star appear out of the darkness that summer evening and scoot across a stretch of sky until it disappeared near the star Vega. (I assume that it was

Vega since that is the brightest star in the summer sky, and this particular shooting star was heading toward that bright star.) By the summer of 1960, as I recovered from a broken arm sustained in a bicycle-riding accident, I really thought that I would like to see the night sky through a telescope—any telescope. Hubble, Webb, and Rubin were not even a gleam in anyone's eye at the time. On the afternoon of September 1, 1960, my uncle stopped by our home and brought out a box containing a bar mitzvah present from him and my parents. Inside the box lay a brand-new telescope, which I subsequently named Echo after the just-launched passive communications satellite. A few hours later, I carefully set the telescope up in our garden. I noticed that between two trees in the southern part of our lawn were two rather bright objects, and I decided to set the telescope up on the brighter of the two. Carefully centering the object in the middle of the finder telescope, I then looked through the eyepiece. I saw what looked like a doughnut of light, complete with a hole in the middle. What was wrong with my new telescope? By playing with it a little, I learned a lot about telescopes in the next few minutes. The most important thing was that by sliding the eyepiece up and down, the doughnut appeared to get either larger or smaller. As I continued to adjust the eyepiece, the doughnut shrunk in size until the hole in the middle disappeared. Almost miraculously, the light in the telescope became the mighty planet Jupiter.

What's more, I saw markings on Jupiter, and three starlike dots nearby that I later learned were Jupiter's moons. Were I growing up in today's culture, the experience from so long ago might have meant nothing. We now have spacecraft that are studying Jupiter at close range, as if we knocked on the front door and Jupiter invited us in. But as amazing as these spacecraft are, nothing can take away from actually looking through a telescope and seeing something in the sky for yourself. In later years, I expanded my collection of telescopes, and I tried to begin the career of each new telescope with a look at my favorite planet, Jupiter.

Both Echo and Jupiter are very special to me. Echo held a place of honor inside my home until I donated it to the Linda Hall Library of Science, Engineering, and Technology in Kansas City. This telescope and the views it provides are still thrilling. And, in 1993, Gene and Carolyn Shoemaker and I discovered a comet which, a year later, slammed directly into that same planet. Now, from our home in Vail, a lovely 14-inch Meade telescope called Voyager shows young people the night sky at a nearby school. It could be someone's favorite telescope. Despite the amazing telescopes that are available now to all of us, it still seems to me that our favorite telescopes are the ones we first looked through, the ones that inspired us to reach for the stars.

—September 2016

CHAPTER 8

GO WEBB!

For a lot of years, it looked as though NASA's latest space tele-scope would never get off the ground. Postponements came with steady predictability. But during the second half of 2021, these delays shrank from months to weeks, and finally days. At last, on Christmas Day the world was treated to the most spectacu-lar launch of the biggest rocket ship since astronauts voyaged to the Moon. I was thrilled beyond words to be able to witness that launch. Over the next few months I waited for the first-light im-ages to appear. They were outstanding. But one particular image, showing the Eta Carinae region in the far southern sky, really touched my soul. It reminded me of the close friendship I had with Bart Bok, who decades earlier had made a career out of studying this wondrous section of sky.

★　★　★

We all got a special and thoroughly delightful present early on Christmas morning. Although I did not set my alarm, Wendee did get up around 5 a.m. I turned on our television set, and what I saw fifteen minutes later was the most thrilling space view since 1969, when Armstrong and Aldrin walked on the Moon. It was

Probably the last direct photograph of the James Webb (Just Wonderful!) Space Telescope, Christmas Day, 2021. Courtesy of NASA.

the spectacular, flawless launch of the James Webb Space Telescope, the start of a mission so perfect and smooth that if Webb could speak, it would have told us that it did not feel any motion whatsoever as it soared away. Even the countdown was unique; it was in French: "Dix, neuf, huit . . ." I did notice a possible hiccup. About ten minutes later, the metal covers designed to protect the telescope during launch fell away while the vehicle was still in powered flight. But a second later, I understood that this was not a hiccup; they were supposed to fall away. The telescope was already out of Earth's atmosphere, and with no air to bother it the protective cover was no longer needed.

As lovely as this experience was for me, the launch was not the most memorable part. That came an hour or so later, when NASA Administrator Bill Nelson gave a speech in which he thanked the many people involved in the process of getting the telescope into space. At the end of his speech, Nelson mentioned a young shepherd boy, sitting out under the stars, looking toward the night sky, and writing a poem about it. That shepherd boy, the administrator went on, went on to become king of Israel.

The poem to which he referred is undoubtedly Psalm 19, the opening four lines of which I quote here, plus an additional one added by nova discoverer Peter Collins, an old friend.

The heavens declare the glory of God.
And the firmament showeth his handiwork.
Day unto day uttereth speech,
And night unto night revealeth knowledge,
[So long as the sky is clear.]

The telescope has now been fully deployed, and it is ready for its final adjustments. Unlike for the shepherd boy, and for all of us on Earth, the sky will always be clear and dark at the Lagrange point 2 (named after Italian mathematician Joseph-Louis Lagrange) in space where the telescope will live. The James Webb, or, as some Canadian friends call it, the Just Wonderful Space Telescope, is there to teach us about the Universe of which we are a part, and I suspect that it will also inspire us to set aside the cares and the news of each day, head out into our backyards, and look up at the night sky.

—February 2022

CHAPTER 9

TELESCOPES NAMED PEGASUS

This article is about much more than a telescope. It is about my grandfather Willie, to whom I was very close. It is about David Zackon of Montreal, Lario Yerino of Kansas City, and Carl Jorgensen, also of Montreal, two of whom have owned and used telescopes they named Pegasus, and about my friend Constantine Papacosmas, who talked my parents into turning Pegasus from a loan into a sale that I still treasure.

My own Pegasus has a lot to answer for. On October 10, 1987, during its first session at a new rooftop observing site, I used it to discover my third comet.

★ ★ ★

In the late summer of 1964, I was leaving the observatory of the Royal Astronomical Society's Montreal Centre with some friends, one of whom was David Zackon. I asked the group if they would like to drop by my house to observe with a 3.5-inch reflector. Before they had a chance to answer, David upped the ante by asking if we'd like to come by his house to look through an 8-inch reflector.

When we arrived at his place, we found a very competent 8-inch reflector with a focal ratio of 7. It gave us fine wide-field views of Jupiter and Saturn plus a few other nice things to see. It was rather pleasant. Just a week later, David telephoned me to invite me for a second look. As we used the telescope to view Saturn, David was adjusting one of the mount's large bolts. As I looked at Saturn I remarked, "I think that's Titan," after seeing one of the planet's large moons. David looked up toward me and said, "No, it is still loose."

Lario's Pegasus.

David told me that he was soon to leave for his university year, and each year he had a tradition of lending the 8-inch to someone who would use it. He then began asking me a few questions, and I told him that I had observed most of the planets, especially Jupiter.

"And the Moon, I suppose."

"Yes. And just a few weeks ago I completed the lunar training program."

"The whole program? All three hundred craters?"

"Yes, and the twenty-six (lettered *A* to *Z*) mountain ranges, valleys, and the Straight Wall."

"You did all this with a 3.5-inch telescope?"

"Yes."

"David, you've just borrowed an 8-inch telescope."

It is difficult to describe the feeling of joy I felt as the new telescope and I returned home and I spent the rest of the night getting acquainted with it. The following day I decided to name it Pegasus, after the large satellites that NASA was launching at the time atop their new Saturn 1 rockets. When my grandfather found out about this a few days later, he was thrilled. "I am especially proud of David [Zackon]," Grandpa Willie said,

Carl Jorgensen's Pegasus at the 2016 AAR.

"for having the insight to know that you would put it to good use."

Over the next several months, Pegasus was used heavily. When David returned from school, Constantine Papacosmas, another good friend, suggested that my parents purchase the telescope for me. David agreed, and we settled on a $400 price for it.

On December 17, 1965, I used Pegasus to begin my comet searching program. Twenty-two years later, on the evening of October 11, 1987, Pegasus and I discovered comet C/1987T1.

The name Pegasus has since been attached to other fine Pegasus telescopes. One of them is a large 20-inch reflector belonging to Lario Yerino from Kansas City. I used this fine telescope one autumn while attending the Heart of America Star Party.

The third Pegasus belongs to Carl Jorgensen, one of my closest friends and someone I have known since 1963. He brought it each year to our Adirondack Astronomy Retreat in the mountains near Lewis, New York. Under the peaceful and beautiful Adirondack sky, when my left eye touches the eyepiece of this telescope, my mind wanders back to those earlier years when I began using my Pegasus during the springtime of my life.

—May 2022

CHAPTER 10

THE JACOB'S STAFF

Even though the telescope is probably the most important invention in modern science, it surely is not the oldest one. On Tycho's island, the great astronomer used a selection of instruments, including several large quadrants, to measure positions of the stars and planets. He'd also used a large armillary sphere, a sextant, and even his first instrument, a Jacob's staff.

The Jacob's staff is one of the most famous astronomical instruments ever invented. It was invented by Rabbi Levi ben Gerson (Gersonides), who lived in France from 1288 to 1344. Despite its age, the Jacob's staff is still in use, mostly in geology. It even found its way into a proposal by the famous geologist Gene Shoemaker, who suggested that it be brought to the Moon aboard one of the Apollo flights.

It also found its way into the arts. It was mentioned in one of Shylock's speeches to Launcelot in act II of The Merchant of Venice.

David with his Jacob's staff.

* * *

By Jacob's staff I swear
I have no mind of feasting forth to-night;
But I will go. Go you before me, sirrah,
Say I will come.

—SHAKESPEARE, *THE MERCHANT OF VENICE*

There are not many references to astronomical instruments in
Shakespeare, even though, in the play's fifteenth-century setting
in Venice, such instruments did exist. One of the more important ones was the Jacob's staff. It was essentially a rod, approxi-
mately a meter in length, with a shorter staff mounted at right
angles to the primary. The observer's eye would be placed at one
end, with the other end at the horizon. By adjusting the distance between the secondary staff and the observer's eye, one
can measure the position of a star or planet relative to the horizon. Also called a *radius astronomicus*, the staff was one of the
most important astronomical instruments prior to the invention
of the telescope. Rabbi Levi gave it a Hebrew name that trans-
lates loosely to "Revealer of Profundities." There actually is no
evidence that the staff was named for the patriarch Jacob, but
it might have been named in honor of Rabbi Levi's friend and
fellow astronomer Jacob ben Makhir.

The story of this remarkable invention, however, is what has
attracted it to me. When I first went to the observatory of the
Montreal Centre of the Royal Astronomical Society of Canada
in October 1960, I met the director of observations, Miss Isabel K. Williamson, who gave me a single assignment: to begin
my astronomical adventure by charting the Moon. Using a map
containing three hundred numbered craters and twenty-six al-
phabetically listed mountain ranges, valleys, and other features,
Miss Williamson asked me to observe all these features and

Crater no. 243, Rabbi
Levi, is the deepest one,
near the center of the
photograph. Courtesy of
NASA.

then make a map of them. I completed the map three years later
during the summer of 1964, and a copy of it is pictured.

As I was beginning my lunar study, I noticed that a version of
my family name was attached to one of the craters, specifically
no. 243, Rabbi Levi. When I excitedly told my family, they merely
took it as information. Years later I finally learned that the Levi
referred to Rabbi Levi ben Gerson, whose grecized (or Greek)
name was Gersonides. I also learned that he was the inventor of
the Jacob's staff. I would imagine that every significant astronomer prior to Galileo must have been familiar with the staff.
Although Tycho Brahe used a sextant to measure the position
of his great supernova of 1572, he doubtlessly could have used a
Jacob's staff for the same thing. There is evidence that variations
of the staff were used as early as the Chaldeans circa 400 BCE.
There is even a sketch of Ptolemy, who lived in second-century
Alexandria during Roman times, holding a version of it.

Although Tycho Brahe replaced the staff with the more advanced sextant, the staff is still used today in geology. Gene
Shoemaker used it while training astronauts for Moon flights

in the 1960s. The staff is still useful in measuring stratigraphic depths of rocks, especially where these depths cannot be easily seen visually.

However famous the Jacob's staff has come to be, it certainly captured the attention of Shakespeare, who was likely aware that it had special meaning to Jewish people, and in fact was portrayed on Israel's stamp collection for the International Year of Astronomy. Little wonder he had Shylock, his major character in *The Merchant of Venice*, swear by it.

—August 2022

PART THREE

SKY LOVERS

Were it not for the many thousands of people, young, middle-aged, and old like me, who are passionate about the night sky, would we have a need for telescopes at all? These days the answer to that is not obvious. With the start of night, telescopes around the world open their circular eyes and peer about. Either they are entirely automated, or their users are in rooms or offices far, far away.

In *Starlight Nights*, Leslie Peltier wrote of how he believed that his telescopes come to life after he has gone to bed. "Sometimes, perhaps in the wee small hours of the night long after I have gone to bed, these two scopes may carry on just such a verbal battle with the 6-inch delivering the final squirrelish blow—'If I cannot see sixteenth-magnitude stars, Neither can you catch a comet.'"

CHAPTER 11

ZECHARIA SITCHIN AND THE DISHARMONY OF WORLDS

A university is a place from which we can see the world not as it is, but as it could be. It is true that, among the boardroom meetings, arguments among the senior staff, and constant variations in the rules required to graduate, ultimately what really counts in scholarship is a cultured ability to think originally and to think for oneself. It may be that some scholastic amateurs see the only difference between popular and scholarly writing is that scholars use footnotes and endnotes. What these critics do not understand is the reason citations are so vital. They are the bricks with which original conclusions are built, one sentence, a single paragraph, at a time.

* * *

Last month, I visited the University of Pittsburgh at Johnstown, Pennsylvania, to give a lecture. The experience was fun and intellectually enlightening. Just before the lecture, I had a good chat with the university president Dr. Jem Spectar, during which he mentioned the name of Zecharia Sitchin, a philosopher who has made some curious predictions relating to astronomy. I decided that Sitchin was worth following up, not because he predicted that some unknown planet might come barging in on Earth someday, but because he was a student of ancient Mesopotamian astronomy. Mesopotamia was a large area between the Tigris and Euphrates Rivers, now a part of Iran. Their astronomy dates back some three thousand years, long before the advent of modern astronomy. Some of our most ancient star names derive from them, particularly a fainter star called Almach. Almach is far from one of the brightest stars in the autumn sky. But it, plus the modern faint constellation of Triangulum, is a part of the ancient Mesopotamian constellation called Mulapin, or the Plough. The earliest writings about the Plough are believed to date back to about 1000 BCE, which would substantially predate Greek astronomy. Where does this leave Sitchin? He has used a rather cursory reading of the ancient texts to propose the idea that a probably fictitious planet called Nibiru is inhabited by a race of extraterrestrials who are responsible for much of the achievements of our Earth. There is absolutely no evidence to back up Sitchin's specific prediction. However, the general idea of planetary collisions, particularly one that took place about 4.5 billion years ago that led to the formation of the Moon, was

Phobos, taken by the Mars Reconnaissance Orbiter, March 23, 2008. Courtesy of NASA.

confirmed by the study of lunar rocks collected by the Apollo astronauts and is now widely accepted by planetary scientists. Ancient literary predictions of modern discoveries are not confined to Mesopotamia.

The famous English author Jonathan Swift, in his *Gulliver's Travels*, imagined a city in the clouds called Laputa, whose scientists had found two moons orbiting Mars. The story was published in 1723, more than one hundred fifty years before the actual discovery of the moons by Asaph Hall on August 17, 1877. He got the sizes right and was close on their distances to Mars. Swift's idea of a cloud city is also found in "The Cloud Minders" episode of the original *Star Trek* series. We may, and should, argue with Zecharia Sitchin. His argument could be called preposterous. But his curiosity about how the ancient people of Mesopotamia pictured the night sky three thousand years ago is not. Had those people not bothered to look up at the night sky and wondered, we would not have walked on the Moon in 1969, or watched a comet impact Jupiter in 1994, or enjoy the exquisite pictures coming down from the new James Webb Space Telescope. Those people from half a world away, and from a distant time, observed the planets and helped to cement the foundation of our modern understanding of the cosmos. This little journey into ancient astronomy was well worth a brief detour with President Spectar, and a look at the sky above the beautiful campus of the University of Pittsburgh at Johnstown.

—January 2018

CHAPTER 12

JOIN YOUR LOCAL ASTRONOMY CLUB

From the idea of what a university means, our book transitions to astronomical observatories, a different kind of institution for higher education. These places include simple personal observatories like Jarnac, to major national observatories like Kitt Peak, and observatories like the Hubble and James Webb Space Telescopes.

Jarnac Observatory, which is our home, is named after the lakeside cottage my grandfather, Willie Levy, built and owned near the village of Ripon, Quebec, in Canada. It is the place from which I observed the Perseids in 1962.

Of the many different astronomy clubs which include me as a member, I have two favorites. One is in the city of my birth, Montreal, Canada. Although my first visit there was in October 1960, I do recall a long and lonely walk across Mount Royal that September in a fruitless attempt to find their observatory. The other club is the Denver Astronomical Society, whose junior astronomical society I joined when I was living in Denver in 1963. This article stresses the idea that an astronomy club can enrich any night-sky experience.

* * *

By a long shot, the best way to get into and enjoy astronomy is to become affiliated with your local astronomy club. Not only do you get access to a ton of knowledge about how to find constellations and to choose and use your first telescope, but also you get a firsthand look at what is happening at the sky from the people who love it the most.

When I was a young teenager, one had to be sixteen years of age to join the society in Montreal. (Thank goodness that rule no longer applies.) But younger people could indeed attend most of the meetings, and on October 8, 1960, I attended my first meeting. Isabel K. Williamson was in charge, and she gave me my first assignment, to create a map of the Moon based on my own observations. Even though I couldn't be a member yet, I embarked on a project that took me three years to complete. (The map is pictured here.) In Canada, most of the astronomy clubs are under the single banner of the Royal Astronomical Society of Canada. There are "centers" within most major Canadian cities. In the United States, the local clubs are independent, and I have been a member of the Tucson Amateur Astronomy Association (TAAA) since 1979, and served as its president from 1980 to 1983. The national association for astronomy clubs is called the Astronomical League.

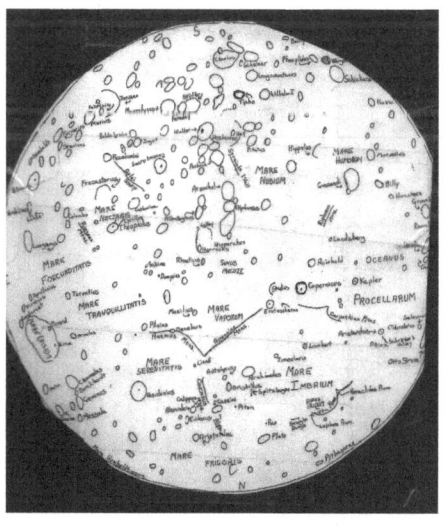

A sketch of the Moon I completed in 1964.

Our local astronomy club has a dark site for observing just east of a tall mountain range.

The observatory that Wendee and I operate from our home is called Jarnac Observatory. Unlike the names of almost everything NASA does, Jarnac is not an acronym. But if it were, Jarnac could be short for Join A Really Neat Astronomy Club. It could also be a vineyard in France, or a board game from decades ago. Or it is the lovely, peaceful place from which I saw a dark sky in August of 1962.

Due to the COVID-19 pandemic, astronomy clubs stopped having in-person meetings. But that hasn't stopped them from indulging in online events. Using platforms like Zoom, Webex by Cisco, or Facebook, online meetings have had an explosion in popularity. They have been so successful that when the pandemic is over, they may continue in some manner.

The most important thing you can get out of an astronomy club is friends. Almost all of my friends are members of one astronomy club or another. They enrich my life and increase my own enjoyment of the night sky a millionfold. I cherish their always welcome insights. In fact, Tim Hunter, one of my closest friends, recently made an independent discovery of a supernova, or exploding star, in the faraway galaxy labelled UGC 10509 and which is hundreds of millions of light years away from us. He may not have been the first to spot it, but his observation has added important new information about the Universe. That star blew up a very long time ago. Its light traveled across space and time until it landed as a speck on one of his pictures, and it is now called Supernova 2020 LQL. This is one of the best things about astronomy. It is an area of study where amateur astronomers can add to our understanding of how the Universe works. Nice work, my friend.

When you next go outside to look at the night sky, enjoy your eyeful of stars. The time after that, try it with your local astronomy club. You couldn't give yourself a better gift.

—August 2020

CHAPTER 13

JEOPARDY JAMES

In this early evening of our lives, Wendee and I loved to relax and watch our television set. Jeopardy! *was one of our favorite programs, and we enjoyed playing along with the various contestants. We particularly looked forward to any astronomy-related topic. In fact, one Final Jeopardy topic had to do with a comet that collided with the planet Jupiter in 1994. "What," went the question, "was comet Shoemaker–Levy 9?"*

★   ★   ★

Of all the programs that Wendee and I enjoy on our television set, the game show *Jeopardy!* is one of our favorites. For a half hour each day, Wendee and I play along as the three contestants try to respond correctly to the host's clues. We were saddened by host Alex Trebek's death last year, but we still enjoy the show whether Mayim Bialik or its top winner, Ken Jennings, is asking the questions. In our tradition, if Wendee or I get a question answered, we applaud each other. It's fun. Last month, the show was unforgettable. In his first thirty-one days as a contestant, James Holzhauer has earned an astonishing $2,462,216 in winnings. On the show that aired Friday, May 31, Holzhauer won $79,633.

Wendee and I particularly enjoy the astronomy clues that come up on shows like *Jeopardy!*. Here is a clue from last Friday: "On November 12, 1833, these meteor showers were seen across all of North America, sparking the serious study of meteor showers." Jeopardy James got it right: "What are the Leonids?!"

The Leonids are a meteor shower which occurs whenever Earth punches its way through the sand-grain-sized debris left by a comet. The debris spreads out across the comet's entire orbit about the Sun. In the case of the Leonids, when the parent comet Temple–Tuttle itself appears in the sky once every thirty-three years, a meteor storm, rather than a shower, sometimes occurs when meteors, or shooting stars, can fall at rates of a meteor per second. It happened in 1833 (the year of the *Jeopardy!* clue), in 1966, and somewhat less intensely over the period from 1996 to 2002.

As I watched this program, my mind harkened back to our visit to Australia in 2001 where we saw 2,406 meteors scratch the sky over the course of a few hours. The display that night began as our group was relaxing on a dry lakebed. A bright shooting star appeared in the east, brightened rapidly as it soared across

Lyrid meteor, April 21, 2020.

Newly rebuilt Shaar building at Jarnac Observatory.

the sky, then disappeared in the west. Before the cheering ended a second meteor repeated the event. At the height of the show, I witnessed *nine meteors appearing simultaneously.* We continued to see meteors well into the morning twilight.

I have observed meteors on more than two hundred nights that began with a night at the original Jarnac cottage north of Montreal. I saw a magnificent, brilliant shooting star low in the southwest. The picture that accompanies this article is of a brilliant Lyrid that appeared to wave at me from the northern sky in late April of this year. Even though I have and use telescopes each night, perhaps my favorite observing session happens when I sit down outside, look up, and watch the sky for these always welcome messages from space that we call meteors. Maybe someday, James Holzhauer will get to enjoy the shooting stars, as well.

—July 2019

CHAPTER 14

GRAVITY

(With Roy L. Bishop)

I have written all these articles myself. Except this one. When I read Roy Bishop's article in the Observer's Handbook *of the Royal Astronomical Society of Canada about gravity, I read his words "Gravity is not a force; it is geometry."—I did not believe him. But when he explained it to me, I finally realized that Roy is the smartest person I have ever known. This article, therefore, was written by both Roy and me.*

★ ★ ★

"Nature had spoken to him."

—ABRAHAM PAIS

Gravity is one of the most fundamental things in physics. Everything and everyone has gravity. The more massive something is, the more gravity it has. When you jump into the air, Earth's gravity brings you back down. What you cannot see while you are in the air is that your gravity brings Earth toward you just a wee little bit, offsetting the extra push away from you that your feet gave Earth when you jumped.

Isaac Newton presented the first ever mathematical description of gravity in 1687. I admit that I know nothing about gravitation, except that it is all around me. I do recall the myth that Newton was sitting under a tree when an apple fell on his head. Supposedly, he then formulated his law of gravity. Did the apple actually fall on his head? I doubt it. But at his childhood home in the village of Woolsthorpe, England, he probably did witness an apple fall from a tree.

During the last half of the nineteenth century, physicists realized that Newton's theory of gravity did not accurately describe the orbit of Mercury, the planet closest to the Sun. Mercury's elongated orbit precesses slightly faster than Newton's theory predicts. Several unsuccessful attempts were made to account for this discrepancy.

Newton's theory, which assumes that gravity is a force, held sway for more than two centuries, until superseded by Albert Einstein's general theory of relativity in 1915. A decade earlier, Einstein realized that mass and energy are two aspects of one thing, and that space and time are interrelated, a blended spacetime. With general relativity, Einstein treated gravity not as a force, but as the geometry of spacetime. The geometry of spacetime is curved by the mass–energy of matter, and the curvature instructs matter how to move.

Now comes the hard part. When Roy Bishop, emeritus professor of physics at Acadia University, pointed out to me that gravitation is geometry, and not a force at all, I didn't believe him at first. But Dr. Bishop is the most brilliant person I have ever had the privilege of knowing. Recently he described gravity this way, and he is right: "Einstein spent several years in an eventually successful attempt to include gravity in a modified description of spacetime. Early in his progress toward that goal Einstein had what he called the happiest thought of his life—that if a person were to fall off the roof of a house, while falling she would not feel a force of gravity. Before she falls, she feels the force of the roof supporting her. When her fall comes to its abrupt halt she feels the ground pushing against her. If she cannot feel a force of gravity while she is falling, why pretend that she felt a force of gravity when the roof supported her before she fell, or that she feels a force of gravity when she is lying on the ground?

"When thinking about the falling lady, Einstein had the fantastic insight that perhaps gravity never was a force. By late in 1915 he had that insight in elegant mathematical form such that the resulting theory, General Relativity, can be used to make precise predictions concerning gravitation."

Einstein was elated when, on November 18, 1915, he found that his general theory of relativity predicted the measured precession of Mercury's orbit. According to his friend and biographer Abraham Pais: "This discovery was, I believe, by far the strongest emotional experience in Einstein's scientific life, perhaps in all his life." Pais then continues with five words that crystallize that profound experience: "Nature had spoken to him." After years of work, on that day Einstein knew that he was the only person on Earth who understood gravity!

Today, there are thousands of people who understand gravity. Roy is one of them. Most of us, including me, are not among them. But reading it described so well is one of the pleasures we can feel as we try to appreciate the wonderful cosmos in which we live. Not only does general relativity correctly predict the precession of Mercury's orbit, but it is essential to the programs used in the GPS navigation system, and it describes the gravitational waves (ripples in the geometry of spacetime) generated by two coalescing black holes, directly detected one hundred years after 1915 by LIGO, the Laser Interferometer Gravitational-Wave Observatory.

—July 2021

CHAPTER 15

ASTRONLINE

This particular entry was written at the height of the COVID-19 pandemic in 2021. Now the pandemic may be entering a less rabid phase, but the idea of online meetings among astronomy enthusiasts continues to surge. I suspect that this form of meeting will be permanent, although it may be supplanted by a slightly different, "hybrid" type of meeting during which both online and in-person sessions are combined.

★　★　★

When the COVID-19 pandemic was at its height in 2020 and 2021, I was busier than ever, meeting many new people, giving lectures, quoting poetry, and advocating observing the night sky.

And Wendee and I have barely left home.

Obviously, I did not give any lectures in person since the COVID-19 pandemic began. On the home front for me, our local Tucson Amateur Astronomy Association meets the first Friday of every month online over the Zoom cloud (see www.tucson astronomy.org). But almost every day, I reconnect with friends in astronomy clubs around the world. On Tuesdays, I am a part of Scott Roberts's weekly Global Star Party. (For more about this, visit https://explorescientificusa.com/products/explore-alliance-global-star-party.) Scott has now had more than one hundred of these wonderful events, and I enjoy each one. On Wednesdays and Saturdays, I am part of the Montreal Centre of the Royal Astronomical Society of Canada, where I meet people I've known for years, especially Carl, one of my best friends since we were teenagers in 1963. As a graduate student at Queen's University in the 1970s, I also was active with the RASC's Kingston Centre. I have also reconnected with the Denver Astronomical Society, a group I joined in 1963 when I was a patient at the Jewish National Home for Asthmatic Children. That experience was precious back then, and it is even more delightful now!

One of the groups, the Warren Astronomical Society in Michigan, does not use Zoom. Instead, they have Webex, which is just as simple to use. I have even participated in sessions sponsored by the Linda Hall Library in Kansas City. Linda Hall is one of the largest science libraries in the world.

Not all of the online sessions are related to astronomy. Our local synagogue has a weekly Torah study session, and Wendee and I are regulars there. They also graciously listen to my poetry quotations, which range from Shakespeare to Chaucer, to this more obscure one (from 1556) from Robert Recorde's *The Castle of Knowledge*:

> If Reasons reach transcend the Skie,
> Why should it then to earth be bound?
> The wit is wronged and led awrie,
> If mind be married to the ground.

When the sessions drag on, as they sometimes do, I can get fatigued since I am not as young as I was in 1963 or 1979. But it is worth the effort, and I sincerely hope that the Zoom/Webex experience will outlive the pandemic when it finally ends. Seeing friends so often like this is wonderful. And on some occasions, I have joined online meetings from a remote site in southeastern Arizona.

Sometimes, my quote tradition includes something from scripture, like this gem from the book of Isaiah:

> Thou stretchest out the heavens as a curtain,
> And spreadeth them out as a tent to dwell in.

My goodness—I never realized how a few words from the Bible could affect me as much as these do. They describe my experience perfectly—outside, I am peering at the curtain of the night sky. Moreover, the observatory out of which I look at the sky, or the observing pad upon which I stand, is the cosmic tent in which I dwell.

—September 2021

CHAPTER 16

THE SPECOLA VATICANA

My fascination with the sky is based as much on faith as it is on science. Science, it seems, is a way of looking at the world based on ideas, experiments, and results that change with time and further experimentation. Faith is quite different. Faith requires a feeling, a sense of belief, and even a modicum of imagination. But I have discovered that a combination of science and faith can be a lifelong, heartwarming experience. I have also learned that both science and faith share a sense of surprise, even joy, when something quite unexpected turns up.

For those of us who have or can develop that sense of faith, the night sky can bring on a whole set of emotions. For me, the emotion I feel most strongly is one of utter peace, and the idea that all is right with the world. I do not often believe that, but I do accept it because that feeling is with me virtually every time I look at the sky.

Having faith does not necessarily relate to a specific religion. I may be Jewish, but I appreciate the idea that so many of the world's great religions have very much in common. For the purpose of this article, the religion I have come to admire is Catholicism. Over the course of my life I have studied the various popes

Oratoire St. Joseph,
a short distance from
my childhood home.

*that have run the church, and they all seemed to share the sense
of peace that is so similar to the sense I feel when I am out under
the stars. I particularly admire Francis, the current pontiff. In
studying the history of the Catholic church, I return to the story of
Galileo. While he was not the first to use a telescope (that honor
probably goes to the Englishman Thomas Hariot), Galileo was
the first to use it to discover Jupiter's four biggest moons. Galileo's
publication of his famous* Dialogue *in 1632 led to his being for-
mally charged and threatened with torture. Galileo was sentenced
to imprisonment, but at the last minute, after frantic pleas, the
sentence was changed to imprisonment at his own home.*

★ ★ ★

The Vatican Observatory, or Specola Vaticana, was born during
the papacy of Leo XIII, around 1891. It was a time when humanity's understanding of the Universe was moving forward, literally
by leaps and bounds. The discovery of a supernova, an exploding
star, within the bounds of the Andromeda galaxy only six years
earlier was proof, beyond measure, that this was a distant galaxy

far away in space, and in time. I like to think there was another, more personal reason for the Catholic church to have an observatory. Until Pope Leo's decision, Vatican astronomy was known for the Gregorian calendar used for astrologic observations for the timing of religious events, and then it became known for the single episode of the Galileo affair. That story began innocuously enough with Galileo's discovery of four moons orbiting Jupiter. In that fateful year of 1610, Galileo instantly became the most famous astronomer in the world. He walked through the papal garden with newly elected Pope Urban VIII, who thoroughly enjoyed talking about the new astronomy with the scientist. And it is said that Urban even encouraged Galileo to publish his *Dialogue Concerning the Two Chief World Systems* (1632). Urban had never realized, however, precisely what the *Dialogue* would actually do. So confident was Galileo in his theoretical argument that he named the character who objected to the new theory Simplicio, and he did make the character out to be somewhat of a fool. Convinced that Simplicio was a parody of the pope himself, Urban was enraged and so upset that he ordered Galileo summoned before the Inquisition and tried for heresy. In a judgment that still reverberates through history, Galileo was sentenced to life in prison. Only frantic pleas led Urban to revise the sentence; Galileo would spend the remainder of his life under house arrest. Even a young John Milton, preparing his magnum opus *Paradise Lost*, was forced to visit Galileo surreptitiously. In this sense, Pope Leo's founding of an astronomical observatory was a sort of penance. Over the years, the institution has accomplished some wonderful science, even building a great telescope atop Mount Graham only 160 kilometers or so east of Tucson. The institution has much of which to be proud. In 1992, Pope John Paul II announced in no uncertain terms that the church erred in its original denunciation of Galileo. It seemed the easiest way to handle this difficult situation. It would have been trickier to attempt a second trial; what if Galileo were to be found guilty a second time? John Paul's approach, simply, directly, and to the point, acknowledged Galileo's incredible contributions to the world as the father of modern science. A few years ago, Wendee and I had a weekly radio program called *Let's Talk Stars.* The show aired on May 18, 2004, and it featured an interview of Galileo himself. In a delightful performance, Brother Guy Consolmagno—now the director of the Vatican Observatory—played Galileo. As I interviewed him, I found myself retreating back into the sands of time, deeper and deeper into a bygone era where people, not unlike us, gazed upward at the stars and asked wondrous questions about the nature of the world.

—December 2017

CHAPTER 17

A WONDERFUL LIFE

It is difficult to overstate the influence that Gene and Carolyn Shoemaker had on my life. From the time I first met them in March 1988, to Carolyn's and my first Palomar observing session together in August 1989, to that stoppage in time and space when our comet Shoemaker–Levy 9 whisked us away from Earth, my time with them marked the pinnacle of my astronomy career. During the course of our observing together, we discovered comet Shoemaker–Levy 9. It seemed then as though the sky itself was working to bring our paths to a closer alignment. And after Gene was killed in a car accident in July 1997, Carolyn decided to continue observing with Wendee and me. Although the observing began to fade as we stopped using film and switched over to electronic observing, our closer friendship never did.

★ ★ ★

One clear evening during the summer of 2019, I was using Pegasus, one of my childhood friend Carl's telescopes, at our annual Adirondack Astronomy Retreat. When my cellphone began to ring, I picked it up with some surprise. At the other end of the line was Carolyn Shoemaker. I was thrilled to hear from her,

From left: Wendee Wallach-Levy, me, and Carolyn Shoemaker at Kitt Peak National Observatory.

as it had been some time since our last contact. Carolyn was doing well, except for a mild loss of hearing. She had called to say that since her daughter and son-in-law had moved to New Mexico, she would be living at the Peaks, a comfortable assisted-living facility in Flagstaff. My colleague Brent Archinal gave me her cell phone number. I was able to speak with her again a few months later. I wanted to find a way to increase the frequency of our conversations. "You speak with your brother Richard every Monday," Wendee commented, and suggested, "Why not call Carolyn every Monday as well?"

For the next eighteen months that's what I did. Carolyn would pick up the phone and announce, "It is David. It must be Monday!" Wendee would often join the discussion as well. But when I called on Monday, August 9, no one answered. After repeated tries, her daughter Linda called to say that Carolyn had had a minor fall and was in the hospital. On Thursday evening, August 12, she went into respiratory arrest. Carolyn died the next morning at 10:40 a.m. Arizona time.

With her husband Gene and the five-year comet and asteroid program we shared, Carolyn was responsible for a very rich period in my life. In fact, virtually every article one reads about the Shoemakers will agree that the discovery and impacts of comet Shoemaker–Levy 9 were the most significant parts of our professional lives.

Carolyn began her observing project a few years after her husband Gene was disqualified as a potential astronaut because of Addison's disease. He decided instead to work on the problem of impacts, not from studying craters as he walked about on the Moon, but from the opposite direction of the comets and asteroids that collide with the Moon, and with Earth. Carolyn quickly learned to become proficient at using the stereomicroscope. She would place two films into the microscope; they were identical except that the second plate would be about forty-five minutes later than the first. The films were almost always identical, except that when an asteroid was moving slowly it would appear to float above the starry background. Carolyn discovered 377 asteroids this way, each one charted until its orbit round the Sun could be determined accurately. When one included the asteroids for which orbits have not yet been determined, that number rose significantly, according to Carolyn, to about eight hundred.

In 1983, Carolyn discovered the first of her thirty-two comets. When their colleague Henry Holt joined the following year, the number of new comets rose rapidly. It was only a year or two after that when she surpassed the number of comets another famous astronomer, Caroline Herschel, discovered, and *Sky and Telescope* published a news note about "Carolyn passing Caroline." I joined the team in 1989. In a sense, passing Herschel's record might have been Carolyn's golden moment, but it wasn't. That came later, on a cloudy and dull day on March 25, 1993. Two nights earlier I had taken two exposures that she was scan-

Shoemaker–Levy 9 discovery photographs. Photographed March 23, 1993.

ning. Suddenly looking up, she announced "I don't know what I have, but it looks like a squashed comet." That was the discovery moment of comet Shoemaker–Levy 9.

Sixteen months later, when the twenty-one pieces of this fragmented comet collided with Jupiter, we got to meet President Clinton and chat amiably with Vice President Gore and share the world's excitement over the first collision of a comet and a planet ever witnessed by humans. It was a satisfying peak to all our careers.

After Gene died in a car accident in Australia, Carolyn continued observing with Wendee and me for several years. One evening she confided that sometimes she wished she had died with Gene. But she did not, and the world was able to enjoy her company for more than twenty-four additional years. The weekly telephone calls began much later, years after Gene had died. I shall miss the deep friendship I enjoyed with Carolyn Shoemaker, the woman whose energy, intelligence, and terrific sense of humor brightened our lives and made the night sky a happier place.

—October 2021

PART FOUR

COMETS

Why are comets so special? I do not really know. My first encounter with a comet might have been a discovery. Sometime in the mid 1950s, while waiting for my Hebrew language class to begin, I looked outside the large window and noticed what looked like a short streak in the sky. Being so young I had no idea what to do about it. The streak did not appear to be moving, so it wasn't an unusual cloud, nor did it appear to be an airplane trail. To this day I have no idea what it was. If the year was 1957, it might have been comet Arend–Roland, or, less likely, comet Mrkos from the same year. I put that memory in storage in my mind. Then in the spring of 1960 I gave a lecture to my sixth-grade class. It was silly to even mention Halley's comet back then, as it had just passed its aphelion, its farthest point from the Sun, and it was not due back for another twenty-six years. Most of all, I had no idea of how that first lecture of mine would affect my life.

Even back then, all those years ago, I did understand that there was something special about comets.
Mr. Powter, our first-rate sixth-grade teacher, intro-
duced us to Shakespeare, and probably quoted for
us Calpurnia's famous line "When beggars die, there
are no comets seen. / The Heavens themselves blaze
forth the death of princes." Thus, I knew that comets
were important. But I could not know then how they
would grow on me. The following chapter suggests
some possibilities.

CHAPTER 18

FAINT FUZZIES

The beauty of even a faint comet, which appears as a fuzzy smudge against the dark sky background, defies attempts at description. Unless a comet gets close enough, and bright enough, to grow a long tail, this is what a comet usually looks like. But each comet brings its own magic, whether it becomes bright enough to outshine the sky's brightest stars, or just remains as a "faint fuzzy." These always welcome visitors from the outskirts of our solar system brighten the darkest of my nights.

★ ★ ★

Long ago, a comet named Palomar (actually known as C [for comet]/2020 T2 [Palomar]) was gliding near one of the most beautiful clusters of stars in the entire sky. It was parading about at about magnitude eleven, which means that for my oldish eyes, it would be too faint to see. In fact, just a few weeks ago I spotted a second comet, named ATLAS. That comet, at ninth magnitude, was so diffuse that I barely spotted it. So, I was not going to try for this other comet.

However, this other comet was named Palomar after one of my favorite observatories! The mighty 200-inch telescope was opened in 1948, just a couple of weeks before I was born, and the big telescope has been sighting stars for more than seventy years. In 1994, I was allowed to sit in the prime focus cage, that beautiful place where light from what the telescope is seeing comes to a perfect focus. So, sighting a comet with that hallowed name would be special. The comet was discovered by Dmitry A. Duev on images taken using Palomar's Oschin Schmidt telescope last October. As the comet was brightening slowly, I learned that on Friday evening, May 14, the comet was planning to glide past Messier 3, one of the brightest globular star clusters in the whole sky.

That was just too much to resist. Clusters of stars are scattered all over the sky, and our own Galaxy has more than a hundred of them. Globular clusters consist of hundreds of thousands of stars. Messier 3 was discovered by Charles Messier, the famous Parisian discoverer of comets; it consists of some half a million stars and is more than thirty-two thousand light years away. At about 11.4 billion years old, it is also one of the oldest things in the Universe. If you want to know more, ask my close friend Peter Jedicke, who probably knows more about globular clusters than anybody else.

With the onset of darkness that Friday evening, I set up my telescope in my backyard observatory and pointed it toward Messier 3. The exquisite star cluster made its appearance. Then I nudged the telescope just a little bit to a nearby field of stars. Suddenly I spotted a faint fuzzy spot precisely where comet Palomar was supposed to be. As I looked around, a meteor scratched the sky to the north. It was a bright and unusual member of the May Ophiuchid meteor shower, a bonus on this unforgettable night.

Comet Palomar is the 219th comet I have seen during my lifetime. Most of these comets have also been faint, barely visible spots of haze. But some have been wondrous. My first comet,

Ikeya–Seki, was the great comet of 1965. Whether a comet is a faint fuzzy of a magnificent comet with a long tail, they are always welcome visitors to Earth's region of the solar system, each one signing, as comet finder Leslie Peltier loved to write, "its sweeping flourish in the guest book of the Sun."

—*June 2021*

CHAPTER 19

OF A COMET AND HISTORY

Ever since I gave a lecture to my sixth-grade class in the spring of 1960, there has been a special place in my heart for comets. I love their appearance, their physics, their history, and most recently, their appearance in literature. Some scholars study the physics and chemistry of comets, while others study their importance to history. Far fewer students delve into how literature treats them, and that is one reason the Hebrew University in Jerusalem allowed me to investigate, amongst other celestial phenomena, the frequent appearances of comets in the works of Shakespeare.

★ ★ ★

Once, I got a good visual observation of comet Tuttle–Giacobini–Kresák, (41P) one of the earliest known periodic comets. It was a fat little "faint fuzzy" spot of light projected against a background of faint stars—nothing to write home about, but for me it was fun just because it was a comet. This comet was only the forty-first that was determined to be periodic when it was rediscovered in 1907, which means that it returns to our part of the solar system again and again. This comet returns every five years or so. However, this comet was actually discovered three

Comet McNaught, an hour before its perihelion, in daylight.

times before the details of its periodic past were finally figured out, and the stories of its findings take us through a good portion of modern history.

This comet was first spotted by the famous comet discoverer Horace W. Tuttle on May 3, 1853, in the little constellation of Leo Minor. It was part of a streak of comets he discovered between then and the first half of the 1860s. Within a few years of this discovery, Tuttle joined the Union army fighting the Civil War. After the end of the war, in 1869 while serving as paymaster aboard the monitor ship *Guard*, he somehow "lost" the considerable sum of $8800 (a very large sum of money at that time) from the accounts of his ship. He was arrested and charged with defrauding the U.S. Navy. At his court martial Tuttle was convicted, but later his sentence was reduced, on approval by President Grant, to a dishonorable discharge from the Navy. One wonders if this semi-pardon had anything to do with his illustrious record as a comet discoverer.

Fast forward through time, to the dawn of the twentieth century when Michel Giacobini was observing from the Observatoire de Nice in France. On June 1, 1907, Giacobini discovered what turned out to be a return of Tuttle's comet. Moving forward again, we arrive at April 24, 1951. I was almost four years old when Lubor Kresák of Czechoslovakia discovered this comet a third time. Now we know that this comet orbits the Sun with a period of precisely 5.419 years. This spring it happened to pass pretty close to Earth, at about a tenth of an astronomical unit (the distance between Earth and the Sun). As we look back at the numerous times this comet was found and found again, we can see how, in 1858, the United States was about to fall into the abyss of its Civil War. In 1907, Lord Baden-Powell was starting the Boy Scout movement. And in 1951, the Korean War was about to begin. As it drifts through the sky, we have the opportunity not just to see a comet sailing through space, but also to take a dip into the ocean of history.

—May 2017

CHAPTER 20

HALF A CENTURY OF COMET HUNTING

December 17 is celebrated as a most important day in my life. More important even than my birthday, it is the never-to-be-forgotten anniversary of the night I began my search for comets.

★ ★ ★

On December 17, 2015, I celebrated the fiftieth anniversary of the start of my search for comets. It has been a long project, with its dawn taking place during the fall of 1965. I was walking toward Westmount High School, in Montreal, Canada, where I was due to sit for a French oral examination; in francophone Quebec, these exams were (and are) required to graduate from high school. I knew that the examiner would ask me what I planned to do as a career.

What did I plan to do as a career? I was interested in astronomy at the time, but I knew that professional astronomy, with all the mathematics involved, was a tall order for me. My thoughts wandered to a comet that had recently been discovered by the Japanese amateur astronomers Ikeya and Seki, and suddenly it hit me: I would search for comets and perhaps even discover one.

By the time I arrived at school, I had planned the entire project, to begin on Friday, December 17, 1965.

At a large table in the Board room at Westmount High, a few hours later, I answered several questions put to me by Mr. Hutchison, the chairman of the French department. Finally, he asked the expected question: what did I plan to do as a career? I smiled and sat straightly and proudly in my chair. "*Monsieur Hutchison*," I replied, "*Je veux découvrir une comète*." Mr. Hutchinson glared at me, and demanded, in English, "How do you expect to make any money doing something like that?" The room dissolved in peals of laughter, and then he continued, "OK, I'll accept your answer. But you'd better discover a comet within twenty years, or else I'll come back and lower your mark!" (I believe he said that because it was the most unusual answer he had ever heard.)

With those sailing orders, I did begin my search with a brief survey between the two brightest stars in Gemini, Pollux and Castor. The search has gone on ever since. I did discover my first comet on November 13, 1984, just one year shy of my French examiner's deadline. In the years since then I have discovered or co-discovered twenty-two additional comets.

I still search for comets. However, much as I enjoy each discovery of a comet, it is the search itself that keeps me going. When I am outside on a clear, dark night, it is just the sky and me, with my telescope as a sort of communications link. On nights like those, I am as close to being in outer space as I ever need to get.

—July 2017

CHAPTER 21

SOME BACKGROUND ON A COMET AND NOVA SEARCH PROGRAM

The second-best decision I made in my entire life was to begin a search program for comets. The best decision I ever made was to marry Wendee. That second-best led to a lifetime of happy hours at the eyepiece, and the first best decision resulted in the happiest years of my life.

* * *

When I started searching for comets and exploding stars called novae, at ten minutes before midnight on the night of December 17, 1965, my program had three aims. (1) To become "very" familiar with the sky through searching for comets and/or novae. (2) To discover either a comet or a nova. (3) To learn as much as possible about comets and/or novae through a research program. As of December 17, 1965, the main interest area was in the field of comets. (4) To observe other comets, too! As of the end of 2015, I have seen 192 different comets. That first goal was really for self-protection, so that the project would be a success even if, as seemed likely, I never discovered a comet. Sure enough, the project helped me learn the sky as never before.

Comet hunting in the predawn with my 16-inch telescope Miranda. I have discovered nine comets with Miranda. Photograph by Wendee Wallach-Levy.

The heart of my program was clearly in the second goal, and I finally reached it on November 13, 1984, with the discovery of my first comet. I'll never forget that night as I looked through my telescope at an object that had never been seen before by *anyone*. I reached that goal a second time early in 1987, and a third time later that same year. On October 2, 2006, I discovered my twenty-second comet, and each of those twenty-two nights will live forever in my memory. Discovering a nova proved a little more challenging, but I did discover independently the great Nova Cygni, just north of the bright summer star Deneb in the Northern Cross, in the late summer of 1975. I was also one of the first to make an independent discovery of Nova Cygni in the fall of 1978, as it erupted near one of my favorite variable stars. Also, I have been involved intensely in studying a cataclysmic variable star, TV Corvi, that my friend Clyde Tombaugh discovered in 1932 based on observations he made on March 23, 1931. On the fifty-ninth anniversary of his observation, I witnessed the

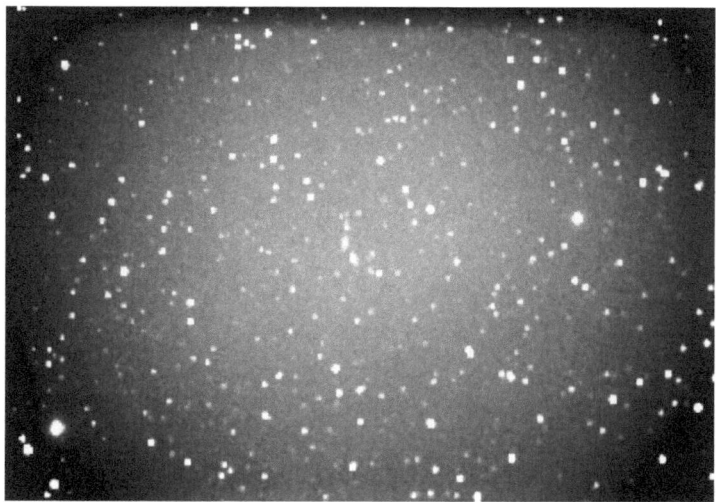

TV Corvi in outburst. Near the top center are three stars in a row. The middle one is Tombaugh's Star.

star erupting again on March 23, 1990. Since then that star has provided one of the great joys of my astronomy life. It erupted yet again on March 23, 2000, and on several other occasions it has erupted near that date.

The date March 23, it turns out, is pivotal in my life. In addition to those March 23 observations of what I call Tombaugh's Star, on March 23, 1992, I wrote a postcard to the young woman who would eventually become my wife. Wendee and I were, in fact, married on March 23, 1997. This turned out to be the fourth anniversary of the day Gene and Carolyn Shoemaker and I discovered comet Shoemaker–Levy 9, the comet that collided with Jupiter. The discovery date of that comet was March 23, 1993.

The idea of a research project connected with the observations was uncertain at first, but it developed very well over fifty years; it was a major portion of both my master's thesis on the comet poetry of Gerard Manley Hopkins and the many comet references and allusions that I found in the writings of William

Shakespeare. These two periods of English literature, along with an additional section about Tennyson, are combined in the second edition of my book *The Starlight Night* (Springer, 2016), my latest attempt to connect the night sky with classic English-language works.

What about the future? While I cannot guarantee that I will be searching for comets until the day I die, I can write that I'm not quite ready to stop the program yet. I will continue to search for comets for a while to come, both visually with my eye at the eyepiece, and with electronic cameras. Even if I never find another comet again, it is the search that remains the most rewarding for me; as the comet hunter Leslie Peltier wrote long ago, "to hunt a speck of moving haze may seem a strange pursuit, but even though we fail the search is still rewarding, for no better way can we come face-to-face, night after night, with such a wealth of riches as old Croesus never dreamed of."

—February 2016

CHAPTER 22

THE DISCOVERY OF COMET SHOEMAKER—LEVY 9

Although I have written much about comet Shoemaker—Levy 9's discovery and subsequent collision with Jupiter, writing about the discovery again after thirty years was a novel experience. The perspective that comes with time gives me a chance to look back on that pivotal time in my life.

★   ★   ★

A lot can happen in thirty years, especially when it involves comets and asteroids that creep across the sky, and even more particularly comets that go bump in the night. Such is the case with comet Shoemaker—Levy 9, which is by far the most important and seminal of the twenty-three comets I have discovered over almost sixty years.

The Jupiter-comet story began for me on September 1, 1960, when I looked through a telescope for the first time. Jupiter was my target and I still recall that view. Years later, Gene Shoemaker proposed that comet Shoemaker—Levy 9 might have been orbiting Jupiter as early as 1929, and that it made a close approach to Jupiter during the year I first sighted the planet. Obviously, I did not see the comet that night; neither did anybody else.

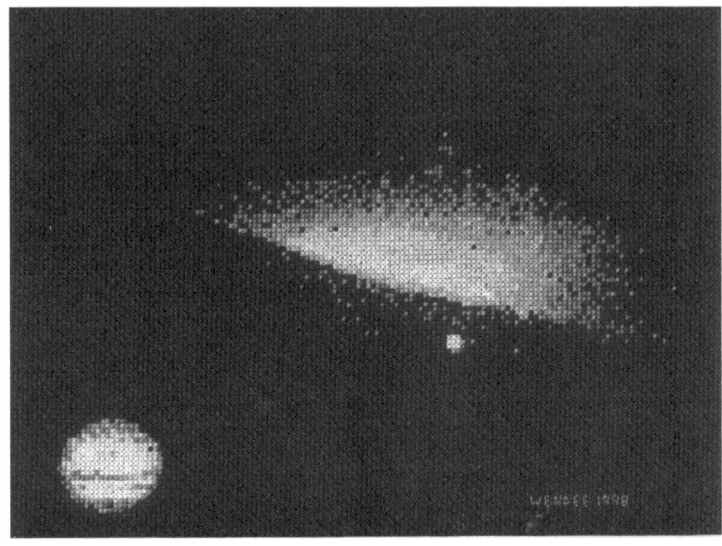

Wendee's artwork on comet Shoemaker—Levy 9 near Jupiter.

On the first night of our March 1993 observing session at the 18-inch Schmidt telescope at Palomar Observatory, Gene Shoemaker developed the first four exposures and found them all blank. It appeared that someone had opened the film box since our February session and exposed the films to light. Examining the pile of films, I suggested that the ones near the bottom might be partially usable. Gene developed one of them and agreed. We continued most of the rest of that night with the partially damaged films until about 3 a.m., when we switched to a new set of prepared films.

On the next night, March 23, I guided an eight-minute exposure. It was difficult to stay centered on the guide star since the glow from nearby Jupiter was interfering. We then did three other fields of sky. Clouds arrived before we had a chance to begin the second set of exposures (so that each field would have two exposures). We stopped observing and left the building to examine the sky. I noticed a slight break in the clouds to the

southwest. Gene teased me as being the "eternal optimist." We had a
strange discussion about money. Gene
pointed out that it costs eight dollars
each time we load a film into that telescope. When I suggested that $8 was
not too much, Gene quipped, "That's
eight American dollars! Not that Canadian play money you try to get away
with!" But after Carolyn agreed that
there was a break coming, Gene said,
"Let's do it!" We somehow managed
to take four exposures before more
clouds came and ended the night.

The 18-inch telescope dome at Palomar
Observatory.

On the afternoon of March 25, the sky was completely cloudy
with snow flurries. Gene was reading *Time* magazine. I was
working on a book about comets. Carolyn was scanning the two
Jupiter films. Suddenly she stopped, looked toward me, and ex-
claimed, "I think I have found a squashed comet." As Gene got
up to look, Carolyn approached me. "You are joking, of course?"
I inquired. Carolyn shook her head. Gene then looked toward
us with the most unusual expression I had ever seen on his face.
Then I looked. There was a long bar of cometary smudge, with
at least five darker centers, each with a tail going towards the top
of its film. There was also a trail of cometary light stretching off
both sides of the central structure.

We needed to get a confirming image. I telephoned my friend
Jim Scotti, who was observing on the 36-inch diameter Space-
watch camera atop Kitt Peak in Arizona. He simply did not believe me when I explained what we had. He said he would try
to find the time to take a confirming picture. Two hours later I
telephoned him again. Jim simply grunted. "The sound you just
heard," he explained, "was me trying to lift my jaw off the floor."
"Do we have a comet?" "Wow, do you guys ever have a comet."

That is the story of how we discovered comet Shoemaker–Levy 9 in the pinnacle moment of our professional lives. Sixteen months later we watched, along with the rest of the world, as the pieces slammed into Jupiter at the incredible velocity of sixty kilometers (thirty-seven miles) per second. (A plane traveling that fast would cross the United States in just over a minute.) We spent time with both then Vice President Al Gore and President Clinton. The whole experience of that wondrous week was unforgettable. And it all began with a single look at Jupiter through my first telescope, a cloudy night, and some damaged film, on the never-to-be-forgotten night of the twenty-third of March, 1993.

—May 2023

CHAPTER 23

OF COMETS, MORE COMETS, AND FRITZ ZWICKY

It is often difficult to bring together studies in history and literature as they relate to comets. But the case of the two comets named for the Zwicky Transit Facility, the program that found them, brings history and literature together all by itself. By observing and charting these two comets, we are invited to explore the life and career of Fritz Zwicky, one of the twentieth century's greatest astronomers. And from history we are gently guided into the literature of poets like Gerard Manley Hopkins, whose gifted insight focused not on the brightest comets, but on a special comet that today could have been included in the Zwicky Transit Facility's bin of comets. While writers normally limit their discussions to the brightest comets, Hopkins explores the not-so-well-studied beauty of a comet before it has a chance to get bright,

> *. . . in some corner seen*
> *Bridging the slender difference of two stars . . .*

This is the only allusion in literature that I have read that deals with a faint comet "scarce worth discovery." These three precious words prove that Hopkins knew the sky, and that he knew comets.

I saw the two ZTF comets, but they were barely visible, scarcely worth discovering. Unless, of course, you know comets like Hopkins understood them. He also understood that as a comet "sights the Sun," she "grows." Hopkins was mostly right, but not all comets grow. If they are on their way out of the inner solar system they will just fade instead of brighten, and some comets, even the promising ones, fall apart as they approach the Sun. The most famous example of such a comet is Kohoutek which reached perihelion early in 1974, and though it did not disintegrate it did not live up to expectations. Moreover, in recent years several comets, instead of growing, did fall apart. But none of these facts diminishes the strength of Hopkins's poetry.

☆ ☆ ☆

Since October 1965, when I spotted my first comet, comet Ikeya–Seki, I have seen more than two hundred thirty different comets. Near the dawn of my passion for the night sky, watching mighty comet Ikeya–Seki rise, apparently right out of the St. Lawrence River, was a sight I shall never forget. The two most recent comets I have seen share the same name; they are both called comet ZTF for Zwicky Transit Facility. This project uses a new camera that offers a very wide field of view. The camera is attached to the large 48-inch Oschin Schmidt telescope at Palomar.

This project has a rich history. It is loosely named for astronomer Fritz Zwicky, one of the founding astronomers at Palomar

Wendee and me in front of Obadiah, our Schmidt telescope, 1996. Photograph by Joan Ellen Rosenthal.

and one of the foremost scientists of the last century. He developed not the big Schmidt but the original, smaller 18-inch Schmidt telescope, the very first telescope atop that mountain. Since this project is named after Zwicky, why are its comets called "ZTF" instead of just Zwicky? It is because the comets are named for the project, not the man.

Dome of the 200-inch reflector at dusk, taken from Zwicky's 18-inch dome.

The historical Zwicky actually had little interest in comets. His career leaned towards the big questions of cosmology, the study of the large-scale issues of the Universe. But he was the first regular user of Palomar's 18-inch Schmidt telescope, the telescope Gene and Carolyn Shoemaker and I used to discover our comets, including the one that collided with Jupiter in 1994. That in itself was a tribute to Zwicky, for it offered insights into how comet impacts contributed to the origin of life on different worlds. Zwicky was not into comets, but he was deeply concerned with the distant explosions of massive stars that he and colleague Walter Baade called supernovae. When he began using the 18-inch there were twelve known supernovae. He discovered 121 supernovae with the 18-inch, 120 by himself and 1 with Paul Wild.

Even though I never met Zwicky, I can share three aspects of him, not including the most famous one in which he called anyone he did not like (please forgive the expression) a "spherical bastard." The expression was intended to mean that no matter from which angle you look, that person is (or was) a bastard. One story I heard from Walter Hass, founder of the Association of Lunar and Planetary Observers, who said that when Zwicky was having a quiet chat in a corridor at Caltech with another astronomer, one could hear him two blocks away. The other involved Zwicky's observing coat, which he left in a closet at the 18-inch observatory building. One night as I was about to observe alone there, as Gene Shoemaker left the building he said, "If you get too cold, you can wear Zwicky's coat!" The thought of that coat haunted me all night. Third, my friend David Rossetter named his large 25-inch-diameter reflector Fritz, after Zwicky's first name. It is a wonderful telescope named for a brilliant man.

In January, the ion or gas tail of comet ZTF showed a sort of disconnection in which the part of the tail closest to the comet was a thin line which suddenly broadened to a larger fan further out. This "disconnection event" was closely tied to a sudden increase in sunspot activity. This ZTF comet teaches us how comets interact with the solar wind.

As this article goes to press, there is not one ZTF comet, but two. David Rossetter and I saw the other one at our club's dark observing site. The second one is much fainter, visible as an amorphous smudge of small, slowly moving haze. As I looked at this second comet, I tried to understand and appreciate the seminal role that Zwicky played in his time. And in our time, that role has expanded to explore in still greater detail the night sky that he loved.

—November 2019

PART FIVE

WHEN WORLDS ALIGN

Writing a fresh introduction to the chapter about eclipses could not have come at a better time, for on Saturday, October 14, 2023, a partial eclipse of the Sun was visible from much of the United States and parts of Canada. My very first eclipse, seen when I was eleven years of age, occurred on October 2, 1959. (On the same date, twenty-eight years later, I discovered my third comet.)

A partial eclipse of the Sun is not one of the best eclipses one can see. Both the 1959 and the 2023 eclipses featured less than a quarter of the Sun being obscured by the Moon. Such is a partial eclipse. That is what a partial eclipse is all about. From a 1 percent eclipse to a 99.9 percent eclipse, the Moon takes a bite out of the Sun. Watching the steady progress of the Moon as it crosses the photosphere of the Sun is the easiest way to detect movement within the solar system. That would happen during an eclipse of the Sun. When the Moon has an eclipse, one can watch as the outer shadow, or the penumbra, of Earth marches across the Moon. At that time, it would not be possible to see the shadow move in "real time," but over the course of ten minutes or so the shadow's movement can be detected.

The idea of seeing motion in the solar system is more than tantalizing: it is beautiful and wondrous. And even though I wrote this article before the 2017 total eclipse of the Sun, I include a rewritten version here because it describes the magic of a total eclipse.

CHAPTER 24

TEACHINGS OF AN ECLIPSE

More than almost anything else, eclipses are teachers. They teach us that the sky is not static. The sky is happening. They teach us what happens when great worlds align. And they teach us about three of these worlds: the Sun, the Moon, and Earth.

★   ★   ★

For many of the readers of this book, the most important thing that happened in the year 2017, particularly for viewers living in the United States, was a total eclipse of the Sun. On August 21, the shadow of the Moon spun across the United States from the coast of Oregon in the morning, crossing the country and pirouetting over the vicinity of Kansas City around noon, and then exiting the east coast of South Carolina late in the afternoon. Almost all of North America experienced a partial eclipse of the Sun. But there is a tremendous, almost indescribable difference between a 99 percent partial eclipse and a 100 percent total eclipse. A 99 percent eclipse is still a partial eclipse, and it takes the extra 1 percent to turn the partial into a total eclipse. If it is only partial, the sky will begin to darken slightly as the Sun's appearance changes from a whole Sun to a crescent. As the

Total eclipse of the Sun, August 1, 2008, from Russia.

Inner corona during a total solar eclipse.

eclipse deepens, the crescent will get progressively thinner until, at the 99 percent level, all that is left is a thin line of sunlight. If you look toward the west, you will see the dark shadow of the Moon approach you, pass by, and then recede as it races to the east. However, the eclipse is still a partial, and then the crescent will widen and brightness will return. That last 1 percent makes all the difference. Those within the path of total eclipse see the Sun's line of light continuing to shrink until all that is left is a point of light. From the west the shadow continues to grow and darken. Looking back at the Sun, you will see what looks like a diamond ring. The diamond is a single bright point of sunlight, and, surrounding the darkened Moon, the Sun's dreamlike corona begins to form. The Sun has vanished, leaving in its place a jeweled crown. The corona, the outer atmosphere of the Sun, is so hot that its temperature can exceed a million degrees Celsius. But as hot as it is, the corona is far thinner than the rest of the Sun; it is almost a vacuum. There may also be erupting prominences coming out of the edges of the Sun. They look like small flames, and most of them are larger than Earth. After a minute or two, the edge of the Moon's shadow approaches, a second diamond ring appears in an outburst of light, and the total phase of the eclipse is over. Just like that.

A total eclipse of the Sun takes our minds off the events of the daily news, which seem so trivial in comparison with the chance alignment of two great spheres in the heavens. The ethereal beauty of a solar eclipse reminds us all that we live on a delicate world that moves around the Sun and that, on rare occasions, our Moon can block out the Sun's light and create a total eclipse, one of the most truly amazing things humanity can witness.

—May 2015

CHAPTER 25

SHADOWS AND AN ECLIPSE

The penumbral eclipses of the Moon are often barely detectable. I observed one on an airplane flying east of Tucson.

★   ★   ★

One lovely clear evening in February 2017, I saw the shadow of Earth extend all the way to the Moon as night fell. Nightfall happens every evening. The Sun sets, and toward the east a dark shadow appears, darkening the sky as it strengthens. After an hour, the "shadow" has spread itself across the whole sky, and it is night. But on February 10, the start of that night was different. Just as Wendee and I saw the first indications of the Earth shadow in the east, the full Moon rose. Only it didn't look full. There appeared to be a shading on the Moon's upper-left portion. What we were seeing was the Earth shadow actually project all the way to the Moon.

It was a lunar eclipse. There are several kinds of eclipses of the Moon and of the Sun. Lunar eclipses can be penumbral, in which the partial shadow of Earth falls on a portion of the Moon. They can be partial, where the full, dark shadow of Earth falls on a portion of the Moon. If the full Earth shadow covers the whole Moon, the eclipse is total. Eclipses of the Sun, which involve the shadow of the Moon reaching a portion of Earth, are different. If the Moon covers a portion of the Sun, then it is a partial eclipse. The full shadow of the Moon tracks along a narrow band, no larger than about 257 kilometers, across a portion of Earth, and along that band there can be a total eclipse of the Sun. That February eclipse was the ninetieth eclipse I have seen. These eclipses range from tiny penumbral lunar eclipses like the one last February, to the grand spectacles of total eclipses of the Sun, of which I have seen ten so far, and I hope to see my eleventh this coming August. But there is more. The night before the lunar eclipse, while I was out in my observatory, I recalled missing one just like this one, decades ago. On January 9, 1963, I was a fourteen-year-old patient at the Jewish National Home for Asthmatic Children in Denver, Colorado. I watched the Moon rise that night during observing session no. 99E, never knowing that a soft penumbral eclipse was actually underway. That early eclipse was a member of a Saros (Greek for cycle), Saros 114. It turns out that, unbeknownst to me, I saw that same eclipse (Saros 114) on January 19, 1981. That eclipse, also a penumbral lunar eclipse, was a repeat of the one I didn't recognize in 1963. The Saros cycle lasts eighteen years, eleven days, and eight hours; and this was the very next repetition of that eclipse. Because of the eight hours (or a third of a day), the eclipse took place at a different time. Eighteen years after that, I missed the next one, because the third of a day meant that the eclipse was visible only in the predawn hours, and I was under a deck of clouds. That brings us to February 10. We were now pretty much back to the same time of day, and the eclipse was much like the one from 1963. This third repetition is called an exeligmos. It is Greek for a period of fifty-four years and thirty-three days. Thus, on February 10, I saw the 1963 eclipse, but fifty-four years later. It will be total along a narrow path that extends from Oregon to South Carolina. From our home in Vail, it will be a deep partial eclipse. The existence of the Saros cycle, and the related exeligmos, makes these wonderful events even more remarkable. This coming August 21, some of us witnessing the solar eclipse might recall seeing the exeligmos one, under similar conditions, fifty-four years ago.

—April 2017

CHAPTER 26

THE TOTAL ECLIPSE OF THE SUN ON APRIL 8, 2024

For this article I have chosen to include the special section, written especially for this book, at the end rather than at the beginning of this writing.

★　★　★

This is a story, not a report on observations.

On April 8, a total eclipse of the Sun tracked across Mexico, the United States, and Canada. Most of the United States enjoyed clear weather, and most of Canada did, too.

We were in Texas. We did not have clear weather.

Admittedly, we knew we might be in for bad luck a week out. But when my friends David and Pam Rossetter came by Friday morning at 5:45 a.m., we knew we would be in for quite an adventure. We arrived at the home in which we planned to stay early Friday evening. It appeared that the house had been vacant for months or years. Although we decided to grin, bear it, and make do, by the next afternoon Scott Roberts, our host, had put us up in a wonderful hotel.

The afternoon before the eclipse, a new report predicted clearing during the eclipse. We were heartened, but that prediction was wrong.

Eclipse day dawned cloudy with drizzle. We arrived at the Explore Scientific site near Leakey, Texas. We did see the Sun for a few seconds now and then. The eclipse began right on time—to the second, even though it may first have been predicted by astrologers in ancient Greece. I remembered how happy Dad was when the 1963 eclipse began the same way. We did get several brief views of the incoming partial. But as the Moon advanced inexorably, the clouds thickened. And as totality neared, it became pretty obvious we would miss the total phase.

About ten minutes before the total phase began, someone in our group asked me to share a poem at the start of totality. The one I had in mind was Ross's speech after Macbeth murders King Duncan:

> By th' clock 'tis day,
> And yet dark night strangles the traveling lamp.
> Is 't night's predominance, or the day's shame,
> That darkness does the face of earth entomb
> When living light should kiss it?

Short and sweet, and so Shakespeare. But two minutes before the onset of the total eclipse, I thought of the closing lines of Wendee's favorite poem, "The Song of Honour" by Ralph Hodgson. I suddenly missed Wendee more than I can write. During the 2017 eclipse my wife opined that she hoped still to be alive to see this one. I understood that, this eclipse, I would have to appreciate it for both of us. The idea of her not being here, at this moment, hit me like a clap of thunder.

The sky was darkening fast. The temperature was falling like a stone. It grew much colder. And still the sky grew darker. It was past noon, and it was night. We were silent.

It was the moment of total eclipse.

I stood and faced the group. I said:

I stood and stared; the sky was lit,
The sky was stars all over it,
I stood, I knew not why,
Without a wish, without a will,
I stood upon that silent hill
And stared into the sky until
My eyes were blind with stars, and still
I stared into the sky.

The group listened with rapt attention. When I was done, there were smiles and some applause. We would not see a total eclipse, but we had a poem. Then there was silence.

Twenty seconds passed.

And then, the Sun appeared in total eclipse. Just like that.

I could not believe it. For about half a minute; for thirty, maybe forty-five seconds, we relished the Sun's corona, the centerpiece of a total eclipse of the Sun. I did not notice the big prominence at the bottom of the Sun, but I did not care. The Sun's corona, circular because this was near the maximum of the sunspot cycle, smiled at us. (At other parts of the cycle, the corona would be more oval.) It was the most dramatic thing I have ever seen.

After that unforgettable, precious sight, clouds came in again. We did get to glimpse the corona on and off a few times after that. I noticed the sky starting to brighten as the end of totality approached. Suddenly it was over.

Only it wasn't.

For one delicious moment the Sun's photosphere appeared. The Sun was shining through valleys at the edge, or the limb, of the Moon. It was a magnificent, stunning view of Baily's beads. First described by Francis Baily after he observed them during the eclipse of May 15, 1836, the effect bears his name. However, the first person to describe this effect was actually Edmond Halley (of comet fame), who recorded them 121 years earlier during

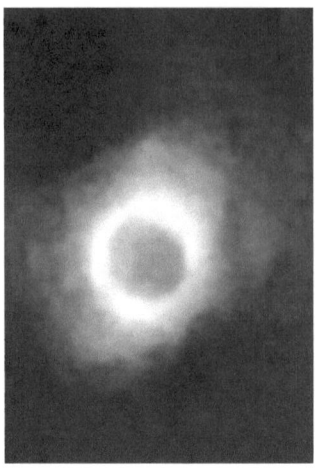

This is what I saw on April 8, 2024, from Leakey, Texas.

the total eclipse of May 3, 1715. What we saw was splendid. And then we got to see a large portion of the ending partial phase. Clouds again obscured the very end of the eclipse.

I sat in my chair, alone. I thought of Wendee. I missed her so much. I could not stop crying. Scott Roberts sat with me and put his hand on my shoulder. Even as I write these words, I am not quite over it.

This eclipse, by far the most dramatic I ever saw, was my twelfth total eclipse, and the 101st eclipse I have seen since October 2, 1959.

* * *

This entry was written just as Star Gazers *was going to press. It tells the story of my view of the total eclipse of the Sun on April 8, 2024, and how the clouds somehow played a game with us. I thought I was getting over the loss of Wendee. Apparently, I am not over it and may never be. Near the end of this book there is an article specifically about Wendee.*

After the eclipse I imagined a scene in which Wendee was seeing the eclipse from Heaven. "God," she said, "Why won't you let Doveed see the eclipse?"

"Well," God replied, "Doveed hasn't been that good a boy lately."

"So what? Let him see the damned eclipse!"

"OK. You're the boss."

After the total phase of the eclipse was over, Wendee said, "Actually, God, you're the boss."

God replied, "Yes, I am the boss. But when it comes to taking care of Doveed, you are."

PART SIX

THE WORLD AND THE SKY

When we simply observe the night sky, we see the Moon, planets, stars, clusters, nebulae, and galaxies. But to grab the whole night-sky experience, we see horizons, houses, roofs, trees, wind, perhaps some clouds, maybe even a canal, and dozens of other things that enhance and add beauty and mood to our observing setting. Relating the night sky to the patterns on Earth, our home planet, defines and enriches our whole night-sky experience. I have felt that experience almost twenty-five thousand times over the journey of my life. From 1956 to the present, each observing session has offered its own singular look and feel. This experience added immensely to the sense of enjoyment and peace that I have felt every time I was out under the stars.

CHAPTER 27

A CANAL, A TELESCOPE, AND A STAR

Ever since Mom and Dad cruised through the Panama Canal years ago, Wendee and I wanted to do the same, and we finally did just that in the fall of 2016. The trip gave us the chance easily to see southern stars, like Achernar, that do not often grace the Arizona sky, and other stars that never climb above our southern horizon.

★ ★ ★

What does a canal have to do with the night sky? For me, plenty. I remember visiting the Lachine Canal many times as a child growing up in Montreal. I even have a dim memory of watching the water raise our boat once. But actually standing aboard the *Norwegian Dream*, a gigantic cruise ship we sailed on to experience the Panama Canal, had to wait until the fall of 2016. As the water surged quietly into and out of the locks on the Pacific and Caribbean sides of the canal, the ship rose and lowered as gently and as quietly as a toy boat in a bathtub. Being part of it was an extraordinary experience.

Being in Panama, on both the Pacific and Caribbean sides, led me to recall another childhood memory. When I was in high

Panama Canal. Miraflores locks near west entrance.

school, I would occasionally bring along a tiny telescope I called Alouette. During recess or lunch, I'd bring the telescope out of the school and get a reading on how many sunspots there were on the Sun. The telescope was so small it didn't capture many sunspots.

I no longer have the original Alouette, but in 1970 I bought a new finderscope. I have now used that telescope, also named Alouette, for forty-six years. Made mostly of war surplus materials, the revised Alouette served as a finderscope, but recently it has evolved into a travel telescope. When I first got it, Acadia University physics professor Roy Bishop helped me get it installed and aligned, so I thought it proper that it be given a long-overdue first-light ceremony. At his Nova Scotia home on the morning of November 7, we used Alouette to enjoy a traditional view of Jupiter, the object I like to use to begin the careers of most of my telescopes.

What does all this have to do with the Panama Canal? I brought Alouette down there and used it to observe stars not normally visible from my Arizona home. In particular, the "star" of the Panama Canal was Achernar. I've seen it from Arizona, but only as it lay sleeping at the horizon, opening its eyes and winking at me briefly before setting again. But in Panama Achernar shone high and prominently in the southern sky.

Because of an effect of Earth's wobble called precession, Achernar appears to be moving northward. In a few thousand years it will become more easily visible from most of the United States and even southern Canada.

Achernar is a big star, 6.7 times more massive and 3,150 times more luminous than our Sun. Even though it is about 139 light years away, it shines as one of the brightest stars in the sky. It rotates about its axis so quickly that it isn't even spherical, but instead it is flattened into an oblate spheroid so dramatic that its equator is half again as fat as its poles. Moreover, Achernar is surrounded by a very large gaseous envelope that grows outward from the star, collapses inward, and then regrows.

It is this final fact of Achernar's envelope that brings me back to the Panama Canal. As I looked through Alouette at Achernar, I could imagine that envelope quietly growing and shrinking, just as the waters in the locks we passed through a few hours earlier rise and fall, lifting and lowering the ships that pass through. The canal helps define two continents. Achernar, even as seen through Alouette, helps define a universe.

—*January 2017*

CHAPTER 28

BASTILLE DAY

Pluto strikes a special chord inside me. I first read about this wonderful planet in one of my children's books, and I heard Clyde Tombaugh, its discoverer, discuss his life and his work in Denver in June of 1963. As I grew older, we became close friends. In his last years he bemoaned the idea that, if Pluto were ever to lose its planetary status, his signature discovery would be taken from him. Clyde died in 1997, and Pluto became a dwarf planet in 2006. I disagree with this for no scientific reason but because I knew Clyde so well.

★　★　★

On Bastille Day, July 14, 1789, the great Parisian prison was set ablaze as a mob, enraged into a fury, lit the spark that set off the French Revolution. Two and a quarter centuries later, on another Bastille Day, a different kind of revolution began as humanity made a great leap forward by sending a small spacecraft past the planet Pluto. On this Bastille Day, the American spacecraft New Horizons zipped by Pluto in humanity's first encounter with that mysterious and famous world. Unlike some of the modern probes that have succeeded in orbiting worlds like the

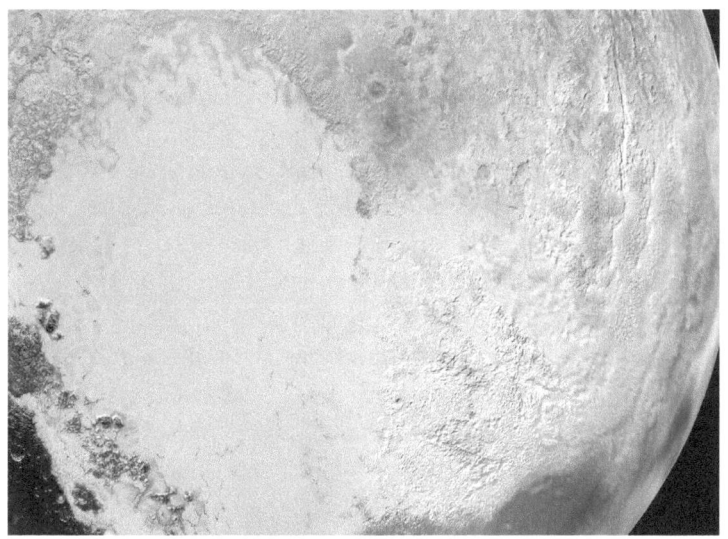

Sputnik Planum on Pluto. Photo courtesy of NASA/New Horizons.

asteroids Vesta and Ceres, and the planet Saturn, New Horizons performed a fast flyby of Pluto. Therefore, its only opportunity to get good pictures spanned a period of slightly more than a day. Before the encounter, the New Horizons spacecraft stopped transmitting anything to Earth and concentrated on its job. Once it was safely past Pluto, it relayed a simple but vital message that it accomplished its tasks.

I couldn't have been at a better spot to enjoy this encounter: the twelfth annual Adirondack Astronomy Retreat at Twin Valleys camp near Lewis, New York. The camp is owned by the State University of New York at Plattsburgh, and the place is fantastic. Early in the morning after the encounter, I was waylaid by some of the amateur astronomers who had studied the biography I had written years earlier of Clyde Tombaugh, the man who discovered Pluto in 1930. They were so fascinated with the simplicity and complexity of the man that they recommended that Wendee and I produce a movie about his life. Much as we appreciated the thought, Wendee and I are not cinematographers, and there we had to let the idea rest.

The day of the flyby of the Pluto system was the cloudiest day of this year's event. Some of the nights had clouds, but we did enjoy several starry nights. A few months ago, the Sky-Watcher company donated a lovely 16-inch-diameter telescope to our retreat, and it saw good use this year. The telescope is named Enterprise after the fictional starship, and several of the participants managed to use it. I was particularly happy that Carl Jorgensen, a close friend for over fifty years, was able to make it this year after he fell on some ice one shivery day last Christmas. He got to use Enterprise several nights, and he enjoyed revisiting the famous double stars, open and globular clusters, and distant galaxies that he always loved to see all these years.

As New Horizons moves ever farther from Pluto into the depths of interplanetary space, we are left also with the best of feelings from a starry sky none of us will soon forget.

—August 2015

PART SEVEN

THE NIGHT SKY EXPERIENCE

This book's penultimate section considers how the sky relates to our lives. In ancient times this was handled through astrology. Today the joining is deeper and, for me at least, far more personal. I may have lost Wendee, but I have not lost my passion for the sky, and I never will.

CHAPTER 29

INNER STARLIGHT

Part of the wonder that touches us during every session of observing, more than the objects themselves, is what we imagine seeing. An object in the sky might evoke a memory of something we have seen long ago, or an imagined thing we might want to see during some future night. Or we may imagine that we are with some of the people we have seen on television programs we enjoy.

Of all television programs I watch, Star Trek *is my favorite. It chronicles the fictitious voyages of the starship* Enterprise. *The writing that follows is from "The Inner Light."*

★ ★ ★

In 1994, *Star Trek: The Next Generation* was one of the most popular shows on television. The episodes were so good that it was easy to tell that the cast was especially enjoying themselves. One of the episodes that year was "The Inner Light." It was a beautiful story in which a strange probe approaches the *Enterprise* and focuses a beam of light on Captain Picard, who loses consciousness and has a dream in which he is living on a distant planet. He enjoys a full life there, with a wife, two children, and a grandson, and he becomes politically active in his community.

He even outlives his wife. One day, his daughter asks him to watch a rocket launch. He hesitates, but then his deceased wife and best friend appear. The Captain then exclaims, "It's the probe that was sent for *me!*"

After enjoying this episode many times, I was reminded of another beautiful story. Written by Nathaniel Hawthorne in 1824, it is called "The Great Stone Face" and concerns a large natural face-like structure hanging near Franconia Notch, across some granite rocks in New Hampshire's White Mountains. The site was magnificent, at least until one night in 2003 when the face fell down in a great big heap. The cliffs are still there, but no more face.

The night sky is much like *Star Trek*, and much like Hawthorne. We look at a group of stars, perhaps a constellation or two, and our brains begin to make connections. On *Star Trek* we share the idea of traveling through space, even if all we have to warp through space with is our two good eyes and a telescope. Some of us may even remember chapter 12 of Hawthorne's masterpiece *The Scarlet Letter* (1850), in which the *A* is likened to a meteor crossing the sky at midnight: ". . . before Mr. Dimmesdale had done speaking, a light gleamed far and wide over all the muffled sky. It was doubtless caused by one of those meteors, which the night-watcher may so often observe burning out to waste, in the vacant regions of the atmosphere. . . . And there stood the minister, with his hand over his heart; and Hester Prynne, with the embroidered letter glimmering on her bosom; and little Pearl, herself a symbol, and the connecting link between those two."

Was the meteor an interpretation of the scarlet *A* parading across the sky? The night sky is full of messages, and only some of those messages come from astronomers. The rest come from people like you and me, people who have innocently stood up and looked at the stars, and who have wondered. The rest come from Shakespeare, and Tennyson, and perhaps even Nathaniel Hawthorne.

The next time you look at the stars, picture yourself not just watching them but reading them. Learn the stories they tell, as interpreted by your favorite writers whether they be Shakespeare, Tennyson, or Hawthorne, or in a Native American or Australian aboriginal legend, or even as thought up by you. What sparks your imagination can be something as simple as a story you have heard, seen, read, or even written. Even in our modern age, the message could indeed be written in the stars.

—January 2019

CHAPTER 30

SHARING THE SKY

If you love the night sky with its stars, planets, and all the other treasures it offers, that is a true joy. But if you spend even a little time sharing that sky with someone younger, the joy is made double. The article that follows describes a teaching and learning foundation that Wendee and I ran for about twenty years. It included many visits to schools and universities, astronomy clubs, and other places where students of the night might gather.

★ ★ ★

On Sunday, April 26, 2015, from 3 to 10 p.m., our National Sharing the Sky Foundation held a public outreach event in Tucson. There were telescopes and related activities set up so that people could observe the Sun safely, and later the night sky that evening. We hoped that our Vail friends would come and enjoy the event with us.

I remember looking up at the sky and *actually seeing* a natural world moving through space. Years ago, in 1983, I watched three comets, IRAS–Araki–Allcock, Sugano–Saigusa–Fujikawa, and Schaumasse, pass relatively close to Earth. These events do not happen very often, so it was a pleasure to sit back, look through my telescope, and enjoy watching the asteroid 2004 BL86 trace a path through the field of my telescope. Seeing an event like that proves to me that space is truly three dimensional, that there are things in the sky that move past us, and that we enjoy watching them as they do.

Have you ever seen a slow-moving meteor burn up in the atmosphere? Before I heard about comets and asteroids doing the same thing, I thought I'd never see anything like them. Fortunately, we are lucky that we can just see them pass by. One day, hopefully very far in the future, an asteroid or a comet will appear, and we'll see it move across the sky. But after an hour of watching it glide by, the speed will pick up as it gets too close for comfort. It is fun to watch a comet pass by, but not so much to watch it collide with us. As it speeds up so that it seems to be going as fast as a passing airplane, it will crack through the top of the atmosphere and with large explosions tear its way until it strikes the ground, digging a crater at least twenty times its size if it hits land. If it hits in the ocean, it will set off a tsunami or tidal wave at least a mile high; the wave will destroy thousands of miles of coastline.

This is not speculation. Earth has seen impacts like this; perhaps the largest one in recent memory was the comet, or maybe it was an asteroid, that hit the eastern shore of the Yucatán Peninsula some sixty-five million years ago. It marked the sudden end of the Cretaceous geological period and resulted in the extinction of 90 percent of all species of life, including all but some of the avian dinosaurs.

We hope this will never happen again, but it will. Hopefully, all the comets and asteroids that come nearby will only pass us by, inviting us to watch them cross our way for a few hours, then recede harmlessly into the vastness of space.

—March 2015

CHAPTER 31

THE NORTHERN LIGHTS HAVE SEEN QUEER SIGHTS

Seeing the northern lights from the Arctic Circle was a dream that came true for me in the winter of 2020. I was a part of the Aurora | 360 experience, which featured an hour-long flight from and back to Whitehorse, the large town in Canada's Yukon territory and the heart and soul of Robert Service's lovely poem "The Cremation of Sam McGee."

★ ★ ★

Ever since I saw my first major display of the northern lights on July 8, 1966, I have been fascinated and delighted by this always welcome show of greenish lights in the sky. But of all the displays I've seen, few can match the thrill of watching them from an airplane cruising high above the Arctic Circle.

In January 2020 I was part of the Aurora | 360 Experience, an event consisting of scientific and cultural presentations surrounding the unique displays of northern lights that can be seen often from the sixty-degree latitude of Whitehorse, in Canada's Yukon territory.

Whitehorse is a fabulous town. It is named after a rock structure on the banks of the Yukon River which resembles the mane of a large white horse. Although it was in use for thousands of years by First Nations cultures, it really got its modern start with the discovery of gold in the Klondike in 1896. The Alaska Highway, built rapidly during the Second World War, passes through Whitehorse.

To me, the city symbolizes two things. One, of course, is the aurora borealis. On the Saturday evening of our trip, we boarded an Air North 737 and took a never-to-be-forgotten flight from Whitehorse to Whitehorse, crossing the Arctic Circle. The sky had been cloudy and very cold, with temperatures hovering around zero degrees Fahrenheit. Looking out of an east-facing window, I spotted a greenish auroral glow the instant the plane cleared the cloud tops. As we headed north the glow brightened rapidly, and soon there were rays, a bright green rayed arc, and dancing green arcs splattered across the sky. As we entered the "auroral oval" just above the Arctic Circle, there was no spot in the sky that was not covered by at least an auroral glow. The northern lights literally surrounded all 360 degrees of the airplane. The three-hour flight was stupefyingly wonderful. I have seen other great auroras, from the big one at the Adirondack Science Camp in 1966, to an even bigger one that covered the whole sky that September, and even a strong red display one night over Tucson, Arizona. But the Aurora 360 | Experience was unique.

What about the other claim to fame of Whitehorse? The city is the centerpiece of one of the most famous poems in all of Canadian history, Robert W. Service's "The Cremation of Sam McGee." It tells the story of Sam McGee, who left his home in Tennessee to join the Klondike gold rush, and who forces the poem's speaker to cremate him if and when he perishes from the cold. There was a real Sam McGee whom Service allegedly met in a bank; McGee's cabin still stands on the grounds of a Whitehorse museum. My father, and most of our family, could quote sections of the poem, but Dad's brother (Uncle Sidney) knew and quoted every word. And when I quoted the final two stanzas

during a lecture at the Yukon Centre of the Royal Astronomical
Society of Canada, several people in the audience quoted them
along with me.

Several days after Sam McGee dies in the poem, an abandoned
boat, the *Alice May*, is used as a makeshift crematorium. As the
flames grow higher, the speaker decides to open the furnace:

. . . Then the door I opened wide.

And there sat Sam, looking cool and calm, in the heart of the
    furnace roar;
And he wore a smile you could see a mile, and he said: "Please close
    that door.
It's fine in here, but I greatly fear you'll let in the cold and storm—
Since I left Plumtree, down in Tennessee, it's the first time I've been
    warm."

There are strange things done in the midnight sun by the men who
    moil for gold;
The Arctic trails have their secret tales that would make your blood
    run cold;
The Northern Lights have seen queer sights, but the queerest they
    ever did see
Was that night on the marge of Lake Lebarge, I cremated Sam
    McGee.

—April 2020

CHAPTER 32

GALAXIES, JUST FOR THE SAKE OF ARGUMENT

This article I just could not resist. The idea of sharing some fun with a friend, particularly one who joins me online almost every week, was just too much to pass up. And now I share it with my readers.

* * *

A few weeks ago, I received a message from Cameron Gillis, an amateur astronomer who wrote that he liked galaxies. Just for fun, I decided to take the opposite approach, a philosophical reversal. The technical term for this behavior is enantiodromia, a term coined by Carl Jung which means the evolution of the unconscious opposite over a period of time. This concept is popular in the yin and yang of Taoism. Thus, if he likes galaxies, then I hate them. As we prepared for our meeting I began to explain the various reasons why I hate them. When, for example, I am observing with a telescope and the Andromeda galaxy enters my field of view, I quickly leave the telescope and ride my bicycle to the end of our driveway and back. The more I stretch the story, the greater the laughter becomes. I especially get annoyed by

Andromeda galaxy, summer 2022.

the dark dust lanes that stretch across its hideous girth. The galaxies in the Virgo Cluster, particularly Messiers 84 and 86, are so bland that I sometimes have to leave the telescope altogether, running screaming into the night.

The worst galaxy is our own. When I look up at the evening sky, the Milky Way obstructs my view as it straddles the night from Cassiopeia all the way down to Sagittarius. The stars are so thick that I can hardly see black sky between them. Except, of course, when I come across Baade's Window. This area of sky rattles me because, there, some darkness appears. Discovered by Walter Baade, this window allows us to see almost to the center of our Galaxy. It is an awful sight. The majesty of the night is nowhere more apparent than when I am viewing the center of our Galaxy, in Scorpius and in Sagittarius, through my telescope. It is wondrous. So wondrous that I still hate it. Because it wastes my time when I am mesmerized by it, the emotion of viewing the Galaxy from my backyard is so strong that it strengthens my heart and pierces my soul.

The worst part of seeing our own Galaxy on a clear autumn night is that the dark lanes of dust straddle its length. Dark areas are called giant molecular clouds. They are not lit by nearby stars; they just are there. In the far distant future, they will generate new systems of stars and planets like our Earth. Giant molecular clouds are often accompanied by hydrogen II (H II) regions in which light from newborn stars excites the hydrogen gas and makes it glow.

In distant external galaxies, dark clouds like these can straddle their whole length. The Andromeda galaxy has several of these dark-cloud regions that one can observe through a small telescope if one looks carefully enough.

Deep in the southern sky, but still visible from most of North America, lies Caroline Herschel's galaxy. It is number 253 in the NGC, the New General Catalogue. Under a bright sky it is hardly anything, but from a dark site it resembles a long, resting caterpillar. It has a most prominent dark dust lane running across its length.

Along with globular star clusters, those round conglomerations of hundreds of thousands of stars that orbit the outskirts of galaxies, including our own, galaxies are the oldest structures in the Universe. The oldest ones started to build within half a million years of the Big Bang, when the Universe was in its infancy.

So much for hating galaxies. When I say that I hate them, I write merely for the sake of argument and humor. Galaxies are almost like people, each one different, each one with its special characteristics. One way of looking at them is to compare their gigantic sizes with our puny selves. But there is another way. Small as we may be, each of us is unique. Galaxies are huge, but aside from their differing shapes, they are still much alike. But in all this Universe, among all these galaxies, there is just one, only one, of each of us. Our ideas, our personalities, are precious.

—November 2021

CHAPTER 33

ON FIRST LOOKING THROUGH BAADE'S WINDOW

The idea of finding a "hole in the sky" with its layer upon layer of obscuring dust, a window that lets us peer almost to the center, must have excited the German astrophysicist Walter Baade beyond expression. He found several such areas, the largest of which is now known as Baade's Window. He found it a few years before I was born. A lifetime later, I still find it magical. However, there is more to the window than just seeing it. It impacts one of my favorite poems, John Keats's fabulous 1816 sonnet "On First Looking into Chapman's Homer." Keats was so excited about discovering Chapman's translation of Homer that he stayed up the rest of the night and composed the sonnet immediately. Likewise, I tried to modify the sonnet to reflect the emotions that Baade must have felt all those decades ago.

★ ★ ★

Much have I travell'd in the realms of gold,
And many goodly stars and clusters seen;
Round all the celestial islands have I been
With eyepiece on telescope to the night sky hold.
Oft of one wide expanse had I been told

That Galileo ruled as his demesne;
Yet did I never breathe its pure serene
Till I heard Baade speak out loud and bold:
Then felt I like some watcher of the skies
When a new planet swims into his ken;
Through his majestic window looks upon the Milky Way
He star'd at the centre of our galaxy.
Like a diamond shining in the sky, with a wild surmise—
Silent, through the mists of space and time.

—JOHN KEATS, "ON FIRST LOOKING INTO CHAPMAN'S
HOMER," FREELY ADAPTED FOR THIS ARTICLE

Lying in the western portion of Sagittarius, the Archer, is a small region of sky that has unusual importance for astronomers around the world and which to me is one of the most beautiful things in the whole sky. It was most thoroughly studied by the German astronomer Walter Baade while using the great 100-inch Hooker reflector at Mount Wilson Observatory in California while searching for the center of the Milky Way galaxy. Before this time, the location of the Milky Way galaxy's center was not well understood.

Walter Baade had an interesting and unusual life. In the mid 1930s, he lost his application papers for United States citizenship. Consequently, in 1941 he was classified as an enemy alien and was held virtually under house arrest. Somehow a compromise was reached, and he was allowed to state his address as Mount Wilson Observatory. With a monopoly of observing time on the great 100-inch telescope, he concentrated his efforts on the Milky Way galaxy.

One of Baade's most important projects was a search for a region of the sky that could be close to the center of the Galaxy. He took good advantage of the wartime blackout over the city of Los Angeles. Intended to help obscure the city from attacking warplanes from Japan, it also darkened the sky significantly so that

Baade could try to find areas near the Galactic Center. Although he did not find much, he did uncover a small area in Sagittarius relatively free of dust. This "window" was slightly south of the main center of the Galaxy. The globular cluster NGC 6522 is at the middle of this area, and NGC 6528 is near its edge. At twelve billion years old, NGC 6522, by the way, may be the oldest globular cluster in our Galaxy.

Astronomers still use this window to study stars in the Milky Way's central bulge. Important information on the internal structure of the Milky Way is still being better understood by measurements made through this "window." The window's shape is irregular in outline and delimits about one degree of the sky, an area of about two Moon diameters. It is centered on NGC 6522. Baade's Window is the largest of the six areas through which stars in the Milky Way's central bulge can be seen. Stars observed through Baade's Window can be called BW (for Baade's Window) stars; similarly, giant stars can be called BW giants. OGLE (Optical Gravitational Lensing Experiment), centered at the University of Warsaw in Poland, and other observation programs have successfully detected extrasolar planets orbiting around stars in this area.

On a rare clear evening during the summer of 2022, I gazed at the clusters and stars through this window. I shall never forget the exquisite majesty of this distant region which, thanks to Walter Baade, allows me to peer toward the middle of the enormous Milky Way galaxy which is our home.

—September 2022

CHAPTER 34

DAFFY DUCK

Our Milky Way galaxy contains so many stars that they seem to appear at random all about the sky. They can form almost infinite patterns, arrangeable into anything we like. But every now and then, something is spotted in the sky that doesn't compute. That was the case with this barely visible nebula outline that resembled the head of a duck that I picked out at an Arizona dark-sky party in 2019.

★　★　★

Agreed, this seems like an awfully daffy title for an astronomy article. But there is method to the madness, and there is a story. During the late summer of 2019, there was a star party in Southeast Arizona that featured a dark sky and five back-to-back perfect nights. As I spent hour after hour hunting for comets, I came across the sprawling North America Nebula in the northern sky constellation of Cygnus the swan. But this time something different appeared. It was a strange structure, the outline of a dark nebula bordered by a slightly brighter cloud. The whole feature was rather subtle, so that sometimes it was there, and then it faded so that sometimes it wasn't. I spent some time trying to de-

The North America Nebula. The Daffy Duck structure may (or may not) be visible on the near left side about a third of the way down. Courtesy of NASA.

termine a name for it. It looked like the head of a duck. I couldn't call it the wild duck nebula, as there is a cluster with that name. And Donald Duck is a bit confusing. So how about calling it the Daffy Duck nebula?

Thus, the structure is named after Daffy Duck. It is number 403 in my catalog of interesting things found during my more than fifty-six years of comet hunting. I believe it is a small dark construction at the northern tip of the North America Nebula, about where Hudson Bay is not accurately located. It could have been where the Gulf of Mexico is, but that area is virtually impossible to spot visually, even under a dark sky. Like the Horse-head Nebula in Orion, it is very difficult to spot and it is best viewed only in a photograph. The accompanying picture shows it at its top, a little to the left of center. The photograph was taken using the Hubble Space Telescope.

There are more than four hundred other celestial objects that have come my way over the years. Beginning with NGC 1931, which I spotted in January 1966, many of these are already well-known deep-sky objects in the night. But a few are interesting groupings of stars, called asterisms, that no one has pointed out before. One of my favorites is a structure of faint stars I call "Wendee's Ring."

These always welcome objects in the sky are fun to observe and they enhance my enjoyment of my hours under the stars. When I can see Daffy Duck, it reminds me of the happy hours I spent as a child at Beaver Lake, an artificial pond near the top of Mount Royal in Montreal, that hosts dozens of mallard ducks. On clear, moonless nights now, I offer a cosmic hello to Daffy Duck and the many objects in the night sky I have come to treasure as good friends.

—December 2021

CHAPTER 35

BACK TO THE MOON

On July 20, 1969, I sat in the auditorium at Camp Minnowbrook on the north shore of New York State's Lake Placid. Together we watched Neil Armstrong and Buzz Aldrin take their first tentative steps upon the surface of the Moon. Four years later, that initial exploration ended with Apollo 17. More than half a century later, human eyes are once again peering toward the Moon in anticipation of a more ambitious second volley of investigation of its mottled and lonely surface.

★ ★ ★

I shouldn't have been surprised by the complete success of the Artemis mission last fall. NASA's *A* team of engineers really know what they are doing. The launch was fun to watch, particularly the brilliant light when its main engines lit up, and it provided some hope that we may actually return to the Moon someday soon.

But somehow, it isn't the same. Something is missing.

For those of us who were alive and young in 1961, we remember President Kennedy's poignant speech to Congress on May 25, 1961, when he asked the nation to commit itself to land-ing a person on the Moon. Only three days after my thirteenth birthday, this was a call I heard distinctly. I did miss the fact that this was the second of three speeches. The first call was during his inaugural address (January 20, 1961): "Let both sides seek to invoke the wonders of science, instead of its terrors. Together let us explore the stars . . ." And at Rice University he gave his third (September 12, 1962): "We choose to go to the Moon."

On August 25 of the summer of 1960, I observed a 99.2 percent partial eclipse of the Moon in which the shadow of Earth covered almost all of the Moon. I remember, a few years later, setting up my first telescope, Echo, across the street to time the Moon passing in front of a star, and explaining to a priest who was passing by that what I was doing might actually assist the Moon mission planning. Or not.

I have already written about where I was on July 20, 1969, during that emotional moonwalk. I listened attentively as the astronauts on Apollo 13 somehow managed to return safely home after the near disaster of their mission. And I watched the interminable countdown hold when, on December 6, 1972, the countdown was stopped just thirty seconds before launch. About two hours later the launch was completely successful, and the program's only geologist, Jack Schmidt, conducted a field excursion two hundred forty thousand miles from Earth, in the Taurus–Littrow valley of the Moon's southern highlands. "I was enormously pleased and proud of Jack," recalled his teacher Gene Shoemaker, "but I was also wistful. There but for a failed adrenal gland, went I." Because of Addison's disease, Shoemaker never made it to the Moon, at least not in life. After he died in 1997, some of his ashes landed on the Moon aboard Lunar Prospector.

In the 1960s, I used the Apollo project to intensify my own passion for observing the Moon through telescopes and binoculars. In 1961, Kennedy set the goal. Eight years later, humans walked on the lunar surface in one of the high points of human civilization. That passion I carry to this day. I still enjoy watching the Moon, looking at its well-known craters and mountain ranges. The Moon is not just a thing in the sky. It is a place. Twelve people have walked across its surface, and with luck more people will someday stroll across its surface.

I will never walk on the Moon. But through my telescope, I shall continue to view the Moon from southern Arizona. And when my eye touches the eyepiece of my telescope, I will be as close to the Moon as I ever hope to get.

—February 2019

PART EIGHT

JOINING THE SKY WITH LIFE

This book's final chapter explores some ideas and concepts that were left out, or needed embellishment, or needed more just plain cogitation than earlier chapters could provide. We explore here some of the connections made between science and faith, and among the many different branches of learning. Some of these divisions reach far from observation to the happy emotion of introducing the sky to a youth, or the incredible feeling of dejection over the loss of someone we love. The sky provides joy for young people, and solace for the grieving.

CHAPTER 36

THE CHRISTMAS STAR

The legend of the Christmas star has always brought generations together, as well as people wanting to find a spiritual or religious meaning to their wanderings under the night sky. It has been a most popular feature in planetarium shows each Christmas. At least one science fiction writer posited the idea that a relatively nearby supernova, appearing as the Christmas star, gave rise to the birth of the Sun, Earth, and the origins of life on Earth. Another possible explanation of what the Christmas star could have been involves a triple conjunction involving Jupiter and Venus, which brought the two worlds so close in the sky that, in the ages before telescopes, they could not be separated. Although the idea of the conjunctions is my favorite explanation, we really do not know what happened so long ago.

★   ★   ★

Said the night wind to the little lamb,
Do you see what I see
Way up in the sky, little lamb,
Do you see what I see

A comet rising in the eastern sky
With a tail as big as a kite
With a tail as big as a kite.

—ADAPTED FROM "DO YOU HEAR WHAT I HEAR?"

BY GLORIA SHAYNE BAKER AND NÖEL REGNEY

This beautiful Christmas carol was written relatively recently, at the height of the Cuban Missile Crisis in 1962, as a plea for peace, a plea that was ultimately successful at that time and which remains so relevant in our time. It went on to become a very popular song, one that I adapted slightly by changing the line "a star, a star, dancing in the night" to "a comet rising in the eastern sky." When Brian Marsden (may he rest in peace) was director of the Central Bureau for Astronomical Telegrams, the world's clearinghouse for anything new that moves or changes in the sky, for example a comet or an exploding star, he and I developed a close friendship. I hoped that one night in late November or December I would discover a comet and report it to him using the words of the song followed by the discovery position, in right ascension and declination, of the new comet. Although I doubt I will ever find another new comet, even if I did, the CBAT is now too automated to allow such informality.

However, this does not mean that I cannot use the famous song to announce the presence of an already discovered comet in the morning sky during this Christmas season. Comet Catalina (C/2013 US10) will form the top of a triangle with bright Venus and fainter Mars in the sky around the morning of December 18. Through a small telescope the comet should be an impressive sight.

It is at this season, where family becomes even more important than usual, that some of us wonder if the Christmas star, made famous in the opening lines of the Gospel according to St. Matthew, was actually a real star or an event in the sky that heralded the birth of Jesus Christ. The relevant portion reads thus:

In the time of King Herod, after Jesus was born in Bethlehem of Judea, wise men from the East came to Jerusalem, asking, "Where is the child who has been born king of the Jews? For we observed his star at its rising, and have come to pay him homage." When King Herod heard this, he was frightened, and all Jerusalem with him; and calling together all the chief priests and scribes of the people, he inquired of them where the Messiah was to be born. They told him, "In Bethlehem of Judea; for so it has been written by the prophet:

'And you, Bethlehem, in the land of Judah,
are by no means least among the rulers of Judah;
for from you shall come a ruler
who is to shepherd my people Israel.'"

Then Herod secretly called for the wise men and learned from them the exact time when the star had appeared. Then he sent them to Bethlehem, saying, "Go and search diligently for the child; and when you have found him, bring me word so that I may also go and pay him homage." When they had heard the king, they set out; and there, ahead of them, went the star that they had seen at its rising, until it stopped over the place where the child was. When they saw that the star had stopped, they were overwhelmed with joy. (Matthew 2:1–10)

Was the star a real event, or was it simply a creation of the Gospel's writer? One early theory was that the star was actually a comet, specifically Halley's comet which made an appearance in the sky of 6 BCE. However, the Magi were known as astrologers rather than astronomers, and surely not modern sky watchers. They might have been more interested in a predicted alignment of planets rather than something real seen in the night sky. With that in mind, there was a highly unusual conjunction of the sky's two brightest planets, Venus and Jupiter, in the evening sky of June 17, 2 BCE, at 6:11 p.m. Venus and Jupiter were so close that evening that, without a telescope (and there were certainly no telescopes at the time), it would have been impossible to see the two planets as anything other than a single bright object. There are problems with this theory. First, the Magi saw a star "at its rising" in the east; the book does not make it clear whether they saw the star from an observing site east of Jerusalem, or in the eastern sky. The planetary conjunction would have been very close to setting, not rising, at the time. Although there are other theories, this is the one I prefer. If it is correct, then it actually suggests a birth date for Christ in the middle of the year 2 BCE; although Christmas comes on December 25 each year, the choice of day actually dates back to the Roman Saturnalia festival just as the days were beginning to get longer in early winter.

Whether you celebrate Christmas or not, this is the season to enjoy family and friends, and to renew ourselves. Let it also be the season, on a clear evening, to head outdoors, look up, and appreciate the sublime majesty of the night sky.

—January 2021

CHAPTER 37

CHRISTMAS AND APOLLO 8

Christmas Eve is a special night. Even the North American Aerospace Defense Command has a tradition of tracking Santa as he delivers gifts to children. The tradition is loosely based on the wonderful editorial written by Francis Church in the New York Sun on September 21, 1897. The final paragraph reads "No Santa Claus! Thank God! He lives now, and he lives forever. A thousand years from now, . . . nay, ten times ten thousand years from now, he will continue to make glad the heart of childhood."

Until 1968, this writing was a key to the season. Then came Christmas Eve 1968.

★ ★ ★

For those of us who were alive back then, where were you on Christmas Eve, in the year 1968? I remember exactly where I was. Sitting in front of my family's television, we were watching a surreal scene on TV. There was a camera peering through a triangular-shaped window on a spacecraft called Apollo 8, out of which was a view of mountains, plains, and craters. And at the bottom of the screen were the words, "Live from the Moon." I have a feeling that most of you, if you were living then, were watching too. The Apollo 8 Christmas Eve broadcast was the most-watched television program in the world up to that time. The announcer on our station, Walter Cronkite, was not saying much. Occasionally he would update us as to what part of the Moon the spacecraft was looking at, but most of the time, the view on the screen said it all. And it was magical.

The year 1968 was a terrible year for the most part. In April, Martin Luther King, Jr., was murdered outside his hotel room in Memphis, and just two months later in Los Angeles Senator Robert Kennedy was assassinated. And two months after that, the Democratic National Convention disintegrated into a riot on the streets of Chicago, with "the whole world watching." That November, Richard Nixon won a close national election. Then came Christmas Eve.

Apollo 8 was not intended to head for the Moon. The Saturn 5 rocket, as tall as a thirty-six-floor building, had never been flown with humans aboard. The NASA picture that accompanies this article, in fact, shows Wernher von Braun, the man who designed the Saturn 5, utterly dwarfed by three engines so large that one could set up housekeeping in each of them. (The other picture is astronaut Bill Anders's epochal "Earthrise.") The Saturn 5's unmanned test flights had been beset by several minor problems, and the Lunar Module, which was intended to land two astronauts on the Moon and return them to the command vehicle, was not yet ready for flight testing. But in August 1968, George Low, manager of the Apollo program office, came up with an ingenious idea:

Wernher von Braun stands next to a Saturn 5 booster rocket showing the huge F1 engines. Courtesy of NASA.

Earthrise photograph, taken by William Anders on Apollo 8, Christmas Day 1968. Courtesy of NASA.

NASA could fly a manned Saturn 5 with only the Command Module. If the launch was successful, it could then proceed to orbit the Moon.

After some debate and a lot of tense moments, Apollo 8 launched on the winter solstice, December 21, 1968. About two hours later, Mission Control radioed a simple message: "Apollo 8: You are go for TLI." After the trans-lunar injection, Apollo 8, with Frank Borman, Jim Lovell, and Bill Anders, was on its way to a Christmas Eve rendezvous with the Moon, and there was nothing left to do but travel and wait.

For me, by far the most memorable part was the astronauts' Christmas Eve message:

We are now approaching lunar sunrise, and for all the people back on Earth, the crew of Apollo 8 has a message that we would like to send to you.

Then each astronaut read from the book of Genesis. Our family was spellbound as we listened to these words. But it was the ending that really turned the year 1968 from one of tragedy to one of promise and hope:

> And from the crew of Apollo 8, we close with good night, good luck, a Merry Christmas—and God bless all of you, all of you on the good Earth.

—January 2016

CHAPTER 38

WHEN POETRY REACHES THE STARS

My interest in the sky might have begun on July 4, 1956, but my delight with the night sky in English poetry began years later, while observing the Lyrid meteors in April 1976. That night sparked a question: Who else might have observed these meteors in years past? Astronomers, writers, musicians, poets? That thought led to my completing a master's thesis at Queen's University in Canada on the poetry of Gerard Manley Hopkins and his passion for the night sky, and twenty-five years later a doctoral dissertation at the Hebrew University in Jerusalem on Shakespeare's special and unique fascination with the night sky.

★   ★   ★

Long, long ago, when I was a student at Acadia University in the Canadian province of Nova Scotia, we studied the poems of Alfred, Lord Tennyson. The English 360 course was taught by one of my favorite professors, Roger Lewis. Tennyson remains one of the truly great English poets, and even in his lifetime he knew that. In 1850, upon the death of William Wordsworth, he was appointed poet laureate by Queen Victoria. In that same year he published *In Memoriam*, arguably his greatest work.

When poetry reaches the stars. A display of the northern lights.

More than a poet, Tennyson enriched his life with a passionate interest in science, particularly the night sky. Did he own a telescope? He surely did. Although he used it often, particularly from his home on the Isle of Wight, he often enjoyed the use of William Dawes's refractor at Cambridge Observatory in England. He viewed some of the great comets of his time, like Donati in 1858 and Tebutt in 1861. He also noticed the discovery of Neptune in 1846. Not only was he aware of these developments, but he also incorporated them into one of the greatest poems ever written, the epic called *In Memoriam*.

In Memoriam grew out of Tennyson's profound loss when his best friend, Arthur Hallam, died suddenly and unexpectedly in 1833 at age twenty-two. His grief evolved into several quatrains of poetry, then many, and he completed the work in 1850. This poem is far more than an elegy. He framed it as a massive commentary on the progress of science during his time, particularly with regard to organic evolution and astronomy. From its dra-

matic opening line "Strong son of God, immortal love," he delves
into what the great telescopes of his time could reveal as

'... Science reaches forth her arms
To feel from world to world, and charms
Her secret from the latest moon?'

We can pass over the poet's wonderful praise of Darwin's theory of evolution and natural selection . . .

Move upward, working out the beast,
And let the ape and tiger die.

We finally encounter the epic's truly wonderful ending. To
write that it is like a bald eagle about to soar in flight is just in-
sufficient. Like a gigantic Saturn 5 as it roars off its launch complex to the Moon, the last two stanzas erupt in a fiery tribute to
creation itself. The poem closes in the Epilogue with a return to
Hallam: "That friend of mine who lives in God" . . .

Tennyson then specifies God as being immortal and loving:

That God, which ever lives and loves

Then he defines the Universe as an ordered realm with a specific goal: "One God, one law, one element." In that single line
Tennyson summarizes how *In Memoriam* explores the interplay
between science and religion. Finally, Tennyson predicts a goal
for the Universe:

and one faroff divine event

In Tennyson's time that goal was not understood. But later,
understanding of Hubble's constant opened the great question as

to whether the Universe will end in a "big crunch" in which the Universe is condensed into a single point as it existed 13.7 billion years ago, or would it continue to expand forever. It is one of these two far-off events "To which the whole creation moves." And thus, we reach the close of *In Memoriam* as it moves proudly among the stars:

That friend of mine, who lives in God,

That God, which ever lives and loves,
    One God, one law, one element,
    And one far-off divine event,
To which the whole creation moves.

—January 2020

CHAPTER 39

GOODBYE, WENDEE

It hurt when I had to write this particular article, and it hurt again when I decided where to place the article in this book. It hurt more still when I discussed the article with Lucretia Free, publisher of the Vail Voice *and the person who, with Wendee, arranged for the column to be written in the first place. About a month after Wendee died, I strolled into our local grocery store to see if the new* Voice *had appeared. It had, and my column about Wendee took up the whole front page.*

★   ★   ★

What follows is the most difficult article I have ever written. On Friday, September 23, 2022, my wife Wendee died. She had been suffering from metastatic breast cancer for over a decade, but this past summer she was truly and clearly suffering. We had an oncologist who was good clinically but who had no bedside manner, and a nurse practitioner who was very good, but a bit of a Pollyanna. Therefore, when Wendee began to destabilize by the hour near the end of September, I was just not prepared for it.

Wendee and I were together for more than thirty years, and we were married for the last twenty-five of them. We got together as the result of a fix-up. When Wendee's mom, Annette Wallach, and my mom, Edith Pailet Levy, resumed their childhood friendship in 1985, my father had just died from Alzheimer's disease. They got together in Montreal and immediately shared stories about their children. Wendee, it turned out, had just separated from her first husband and I was long since divorced from my "practice wife." They decided to try to bring us together. Wendee was the first to reject the idea: "I am a dog person; he is a cat person," she said; "I am an athlete, and he is a couch potato." (I could say that over time I became a dog person and Wendee became a couch potato, but I won't.)

I just ignored my mom's suggestion. Every year or two, Mom would repeat her idea. After seven years, Mom asked again, and when I still had not done anything she annoyingly chastised me and said, "Forget the whole thing. Forget I ever asked you." That was a challenge. On March 23, 1992 (one year to the day before I took the two photographs that led to the discovery of comet Shoemaker–Levy 9 that would strike Jupiter in 1994), I typed out a postcard to her in Las Cruces. She replied, and we finally met that summer. After lunch and a conversation with her and her two sisters, I returned to Clyde and Patsy Tombaugh's house, where I was staying. When Patsy answered the door she asked, "Well, how did your date go?" I looked at her and replied, "Patsy, I have just spent time with the three most beautiful women I have ever met!"

Early in our relationship, we were driving near Las Cruces. It was a clear dark night and we got out of the car. Wendee looked up and asked me, "What star is that?"

"That bright star," I answered her, "is Vega." Just then Wendee recalled that her first husband, long since divorced, had warned her that he would never answer her questions more than once. Wendee then inquired of me, "David, if I were to ask you every night, looking at that same star, the same question, 'What star is that?,' what would you do?"

"I would explain to you, every night," I replied, "that that star is Vega. And I would never, ever, tire of it."

On another evening, I was driving Clyde Tombaugh, who discovered Pluto, and his wife Patsy, back from a dinner engagement. Clyde was sitting up front with me, and Wendee was in the back seat with Patsy. "Clyde, I am going to take you home first and then I will take Wendee home."

"David, why not just drop Wendee off on the way? It would be faster."

"Clyde, I may want to hug her and give her a big kiss."

"That's okay. We'll wait!"

The group in the car got silent. I looked back toward Wendee, then to Clyde, and I said, "Clyde, I am taking you and Patsy home first." As Wendee and Patsy giggled in the back seat, Clyde said, "OK. Now that you explain it!"

Wendee and I were married in the Flandrau Science Center on March 23 (that magic date again), 1997. The reception at our home featured comet Hale–Bopp and a lunar eclipse. Our first few months were difficult. Gene Shoemaker was killed in a car accident in Australia that July, and I had two cancer surgeries (prostate and right kidney) later that year. But as I recovered, our marriage became fun and interesting. We traveled everywhere. Possibly her favorite trip was to the outback near Alice Springs, where we observed over two thousand meteors on a single night in November 2001. Because for a short period we saw one meteor per second, I considered the Leonids that year a meteor storm.

We took three trips to Israel together, the last two of which were part of my doctoral work at the Hebrew University on the night sky in Shakespeare's time. I loved that particular period in my life, and Wendee and I had a lot of fun navigating the multitude of rules and regulations that the university appeared to make up as we went along. Near the end of that process, I wrote a routine question about the dimensions of the European paper

I needed to use. The next morning, I found Wendee looking at her email. "I need for you to read this message now, right now," she said. "Is it good news?"

"I do think so." The letter was from the Hebrew University, announcing that the University Senate had just awarded my PhD and that they hoped we would come to Israel to receive the degree in person. We spent the remainder of that happy day making flight arrangements.

Wendee served as director of our Jarnac Observatory, and I served as her assistant. I used it every clear night. During the twenty-six years we lived in our Vail home, I discovered only one comet, in 2006. The comet was confirmed by the Central Bureau for Astronomical Telegrams just as we returned home from the Yom Kippur services. I was so overwhelmed by the message that I printed it, and then, without a word, cried as I walked back to the house and showed it to Wendee.

Wendee's greatest joy was not so much me, but our daughter Nanette and our grandchildren Summer and Matthew. One night, when she was a college student, our granddaughter Summer contacted us to inquire of a bright red star she noticed high in the southern sky. What followed was a wonderful conversation about Mars, Percival Lowell, and the possibilities of life somewhere on that distant world. Our grandson Matthew provided us with golden opportunities to show how, when he looks at the stars, he can escape the chatter of the nightly news and appreciate the big picture of the night sky and the Universe at our doorstep.

Our marriage gave me an opening to write some books, but my favorite book began when one morning I found Wendee reading intently. "You never told me you wrote a book about your dog when you were ten years old." She found that crazy old book the most delightful she had seen. She even quipped that all my later books went downhill from this one. (We enjoyed a good laugh over her comment.) She wanted me to revise it.

By this time, around 2013, she had received her diagnosis of breast cancer. For years she did very well, until the end of last summer when she needed surgery to remove her ovaries. After that she began chemotherapy, which worked for a few months.

By the spring of 2022, all the treatments stopped working, as though her body had decided that she had had enough; her time was up. Wendee insisted that I go to that year's Adirondack Astronomy Retreat, but she was obviously suffering. We made a 911 call in mid September, and a second one two weeks later. In between I presented her with the first copy of *Clipper*, my new book for children (RJI Publishing, 2022). She was able briefly to hold it up and examine the front and back covers. With that second 911 call, I was pretty certain she would never be coming home. Wendee died on Friday evening, September 23. She was seventy-three years old.

The night before her funeral, our son-in-law Mark, our grandson Matthew, and I were enjoying an evening in the observatory. Matthew saw a bright meteor, and as I questioned him on its direction I saw a faint one. Mark saw a third meteor. I like to think that this minor outburst of the October Cygnid meteor shower—three meteors within a period of about five minutes—was Wendee's goodbye. Rest in peace, my sweet Wendee.

—November 2022

EPILOGUE

The Soul of a Library

This book's final article touches the heart of what life could be about.

When I was very young, Mom criticized the fact that neither my brother Gerry nor I read enough. There is almost nothing more that I love to do now but read. In high school we were assigned a book of our choice to read and then to write a report on it. The book I chose had just arrived at the library of the Royal Astronomical Society of Canada's Montreal Centre. It was Leslie Peltier's autobiography Starlight Nights: The Adventures of a Star-Gazer *(Harper and Row, 1965), and it remains the finest work of literature I have ever read. I planned to read a chapter a day but once I got through the opening paragraph I could not put it down. There are many books that tell you how to watch the sky. This one tells you why. This book captures the soul of the sky. Years ago, I suggested that Dad would enjoy reading* Starlight Nights. *To write that he did is an understatement. "I've always appreciated your single-minded devotion to the night sky," he said to me. "Now I understand why."*

Dad always loved to read, and he particularly enjoyed science fiction. He once suggested that his love of reading led to his son's interest in the night sky. But when, during the summer of 1960, I became totally enthralled with it, he suggested I was going too far. "We talked about astronomy last week," he offered. "And we'll talk about astronomy next week. But this week, we shall discuss something else. Finally," he went on, "don't make astronomy the most important thing in your life." "You are right," I thought. "I won't. Instead, I'll make it the only thing in my life."

As the years passed and I hopefully matured a bit, Dad changed his tune, especially when, in his final months of life, I could tell him that I had discovered a comet. Comet Levy–Rudenko, found on November 13, 1984, would be the first of a twenty-three-comet run that included thirteen Shoemaker–Levy comets plus one that Tom Glinos recorded on his telescope, named Hyperion, and which was credited to Jarnac Observatory. Dad was not around long enough to see any of my subsequent discoveries, but Mom certainly was. When my own life got incredibly busy in the wake of my discovery, with Gene and Carolyn Shoemaker, of comet Shoemaker–Levy 9, I asked Mom if she would like to accompany us to Washington. Her one-word answer: "Yes!"

I think both Mom and Dad would have been pleased that I completed, at last, my doctoral dissertation. "Finish what you start" was a maxim both Mom and Wendee agreed with. In the years since then, I have accumulated a large collection of poetry (and some prose) that involves the night sky. I read a verse or two at each meeting I attend, and with it my understanding of the warm relation between reading the words and reading the sky becomes more fun, more satisfying, and more fulfilled.

★ ★ ★

There, in the night, where none can spy
All in my hunter's camp I lie,
And play at books that I have read
Till it is time to go to bed.

These are the hills, these are the winds,
These are my starry solitudes;
And there the river by whose brink
The roaring lions come to drink.

I see the others far away
As if in firelit camp they lay
And I, like to an Indian scout,
Around their party prowled about.

So, when my nurse comes in for me,
Home I return across the sea
And go to bed with backward looks
At my dear land of Story-books.

—ROBERT LOUIS STEVENSON, "THE LAND OF STORY-BOOKS,"

Just one day after Earth passed its closest point to the Sun in its orbit, its perihelion, the American Astronomical Society was having its annual meeting online, the United States Congress was validating the results of the 2020 national election, and Wendee and I were settling in for a civics lesson about the way the United States government works. The day turned out quite differently.

Shortly before noon, on our television set a news ticker appeared. It announced that two buildings in the Library of Congress (LOC), the James Madison, and quickly afterwards the Adams and Jefferson buildings, were being evacuated. That news sent a chill through me. The LOC is one of the finest libraries in the entire world. It contains more than 170 million books, of which more than thirty are books I wrote entirely, partially, or at least a foreword to. It also includes all of the more than two hundred "Star Trails" columns I wrote for *Sky and Telescope* magazine between 1988 and 2008, and dozens more I wrote for other magazines and journals. Only the British Library, with over two hundred million books, is larger than the Library of Congress.

This event was personal for me. A few minutes later, when the entire Capitol complex was stormed, it was personal for all of us. All of us had reactions to this, but besides the feelings I shared with most of you, I had an additional feeling—specifically about the library.

How many books does it take to make a library? When I was a child in 1963, a teacher gave the best answer I've ever heard: "two books." For me, a library—any library—is every bit as priceless as a dark sky. The wisdom of the ages is contained in each library, from the LOC to a child's collection. I have never gone into a library without feeling better when I exited. The idea that this magnificent collection was threatened that day was terrifying. It was also a day that offered a glimmer of hope.

I have read many books over my lifetime, from *The Cat in the Hat* to my boxed set of *The Lord of the Rings*. One small treasure, Jene Lyon's Golden Book *Our Sun and the Worlds Around It* (Golden Press, 1957), began a lifetime of stargazing. That gem, by the way, also lives in the Library of Congress. What is more, I have never encountered a really bad book. When an author places her or his thoughts on paper in a book, that book immortalizes those thoughts.

My glimmer of hope is that Capitol Hill and the Library of Congress are never threatened again. They belong to We the People, and stand beautifully in Washington, D.C., to govern us, teach us, and encourage us to follow our dreams and reach for the stars.

—April 2021

NOTES

1. Star Gazers

upon its frame, and pointed to the sky: Wordsworth, William, "Star-Gazers," stanza 1.

One after one they take their turns, nor have I one espied: Wordsworth, stanza 8.

6. Getting Loose Change

When beggars die, there are no comets seen: Shakespeare, William, *The Tragedy of Julius Caesar*, 2.2.30 – 31.

Part Three: Sky Lovers

Sometimes, perhaps in the wee small hours: Peltier, Leslie, *Starlight Nights: The Adventures of a Star-gazer* (New York: Harper and Row, 1965), 4.

14. Gravity

Gravity is not a force; it is geometry: Bishop, Roy, "Orbital Motion," in *The Observer's Handbook*, ed. James Edgar (Toronto: Royal Astronomical Society of Canada, 2021), 27.

This discovery was, I believe: Pais, Abraham, *Subtle Is the Lord* (Oxford: Oxford University Press, 1982), 253.

Nature had spoken to him: Pais, 253.

15. Astronline

If Reasons reach transcend the Skie: Recorde, Robert, *The Castle of Knowledge* (1556), preface.

Thou stretchest out the heavens as a curtain: Book of Isaiah 40:22.

Part Four: Comets

When beggars die, there are no comets seen: Shakespeare, William, *The Tragedy of Julius Caesar*, 2.2.30–31.

18. Faint Fuzzies

its sweeping flourish in the guest book of the Sun: Peltier, Leonard, *Starlight Nights: The Adventures of a Star-gazer* (New York: Harper and Row, 1965), 43.

21. Some Background on a Comet and Nova Search Program

to hunt a speck of moving haze: Peltier, 231.

23. Of Comets, More Comets, and Fritz Zwicky

in some corner seen: Hopkins, Gerard Manley, "I am like a slip of a comet," lines 2–3.

scarce worth discovery: Hopkins, line 2.

sights the Sun . . . grows: Hopkins, line 6.

26. The Total Eclipse of the Sun on April 8, 2024

By th' clock 'tis day: Shakespeare, William, *The Tragedy of Macbeth*, 2.4.6–10.

I stood and stared; the sky was lit: Hodgson, Ralph, "The Song of Honour," lines 208–215.

31. The Northern Lights Have Seen Queer Sights

Then the door I opened wide: Service, Robert W., "The Cremation of Sam McGee," lines 56–68.

35. Back to the Moon

I was enormously pleased and proud of Jack: Paraphrase of Gene Shoemaker, quoted in Levy, David H., *Shoemaker by Levy: The Man Who Made an Impact* (Princeton: Princeton University Press, 2000), 264.

37. Christmas and Apollo 8

Apollo 8: You are go for TLI: Mission Control, *Apollo 8 Technical Air-to-ground Voice Transcription*, time stamp 00:02:27:22, https://history collection.jsc.nasa.gov/JSCHistoryPortal/history/mission_trans/AS08 _TEC.PDF.

We are now approaching lunar sunrise: Anders, Bill, *Apollo 8 Technical Air-to-ground Voice Transcription*, time stamp 03:14:04:53.

And from the crew of Apollo 8: Borman, Frank, *Apollo 8 Technical Air-to-ground Voice Transcription*, time stamp 03:14:08:07.

38. When Poetry Reaches the Stars

Science reaches forth her arms: Tennyson, Alfred Lord, *In Memoriam*, part XXI, stanza 5.

Move upward, working out the beast: Tennyson, part CXVIII, stanza 7.

That friend of mine, who lives in God: Tennyson, epilogue, stanzas 35–46.

INDEX

ABOUT THE AUTHOR

David H. Levy is arguably one of the most enthusiastic and famous amateur astronomers of our time. Although he has never taken a class in astronomy, he has written over three dozen books, has written for three astronomy magazines, and has appeared on television programs featured on the Discovery and the Science channels. Among David's accomplishments are twenty-three comet discoveries, the most famous being Shoemaker–Levy 9 that collided with Jupiter in 1994, a few hundred shared asteroid discoveries, an Emmy for the documentary *Three Minutes to Impact*, five honorary doctorates in science, and a PhD from the Hebrew University of Jerusalem (2010), which combines astronomy and English literature.